Does Technology Drive History?
The Dilemma of Technological Determinism

edited by
Merritt Roe Smith
and
Leo Marx

The MIT Press
Cambridge, Massachusetts
London, England

Set in Baskerville by DEKR Corporation.
Printed and bound in the United States of America.

Library of Congress Cataloging-in-Publication Data

Does technology drive history? : the dilemma of technological
determinism / edited by Merritt Roe Smith and Leo Marx.
p. cm.
Includes bibliographical references and index.
ISBN 0-262-19347-7.—ISBN 0-262-69167-1 (pbk.)
1. Technology—Social aspects. I. Smith, Merritt Roe, 1940–.
II. Marx, Leo, 1919–.
T14.5.D64 1994
303.48′3—dc20 93-21422
CIP

10 9 8 7 6

To the memory of Dr. Bern Dibner and with thanks to David and Frances Dibner for all they have done to encourage and support the study of the history of science and technology.

Contents

Introduction

Leo Marx and Merritt Roe Smith

If "determinism" sounds "cold and mathematical," as William James once noted,[1] then "technological determinism" surely sounds even more forbidding. Yet our subject is not nearly as esoteric as that arcane name suggests. By now, most people in modernized societies have become habituated to the seeming power of advancing technology (and its products) to change the way they live. For them, indeed, the steady growth of that power is just another self-evident feature of modern life, an obvious fact that calls for no more comment than the human penchant for breathing. As an explicit idea "technological determinism" may mean nothing to them, but the phenomenon it ostensibly represents is an omnipresent aspect of their awareness.

A sense of technology's power as a crucial agent of change has a prominent place in the culture of modernity. It belongs to the body of widely shared tacit knowledge that is more likely to be acquired by direct experience than by the transmittal of explicit ideas. Anyone who has witnessed the advent of the computer, for example, knows a great deal about how new technology can alter the very texture of daily life, and has gained this understanding as more than a bystander. Even those who do not use computers have had to accommodate their ways to some of its requirements in supermarkets, post offices, banks, libraries, schools, airlines, hospitals, or the military services—few departments of contemporary life remain unaffected by the new information technology. But of course the computer is only one of the radically new science-based technologies—along with television, jet aircraft, nuclear weaponry, antibiotics, the contraceptive

1. "The Dilemma of Determinism," in *Essays in Pragmatism* (Hafner, 1951), p. 40.

pill, organ transplants, and biogenetic engineering—whose transformative power has been experienced by millions alive today. For some three centuries, direct firsthand experience of that power has been a well-nigh universal feature of life in developed and developing countries.

The collective memory of Western culture is well stocked with lore on this theme. The role of the mechanic arts as the initiating agent of change pervades the received popular version of modern history. It is embodied in a series of exemplary episodes, or mini-fables, with a simple yet highly plausible before-and-after narrative structure. Before the fifteenth century, for example, Europeans are said to have known little or nothing about the western hemisphere; after the compass and other navigational instruments became available, however, Columbus and his fellow explorers were able to cross the Atlantic, and the colonization of the New World quickly followed. Newly invented navigational equipment is thus made to seem a necessary precondition, or "cause," of—as if it had made possible—Europe's colonization of much of the world.

Similarly, the printing press is depicted as a virtual cause of the Reformation. Before it was invented, few people other than the clergy owned copies of the Bible; after Gutenberg, however, many individual communicants were able to gain direct, personal access to the word of God, on which the Reformation thrived. As a final example, take the story, favored by writers of American history textbooks, about the alleged link between the cotton gin and the Civil War. In the late eighteenth century, slavery was becoming unprofitable in the American states; but after Eli Whitney's clever invention, the use of African slaves to harvest cotton became lucrative, the reinvigorated slavery system expanded, and the eventual result was a bloody civil war.

The structure of such popular narratives conveys a vivid sense of the efficacy of technology as a driving force of history: a technical innovation suddenly appears and causes important things to happen. It is noteworthy that these mini-fables direct attention to the consequences rather than the genesis of inventions. Whether the new device seems to come out of nowhere, like some *deus ex machina,* or from the brain of a genius like Gutenberg or Whitney, the usual emphasis is on the material artifact and the changes it presumably effects. In these episodes, indeed, technology is conceived in almost exclusively artifactual terms, and its materiality serves to reinforce a tangible sense of its decisive role in history. Unlike other, more abstract forces to which historians often assign determinative power (for example, socio-economic, political, cultural, and ideological formations), the thingness or tangibility of mechanical devices—their accessibility via sense perception—helps to create a sense of causal efficacy made visible. Taken together, these before-and-after narratives give credence to the idea of "technology" as an independent entity, a virtually autonomous agent of change.

Today a similar idea informs the popular discourse of technological determinism. It is typified by sentences in which "technology," or a surrogate like "the machine," is made the subject of an active predicate: "The automobile created suburbia." "The atomic bomb divested Congress of its power to declare war." "The mechanical cotton-picker set off the migration of southern black farm workers to northern cities." "The robots put the riveters out of work." "The Pill produced a sexual revolution." In each case, a complex event is made to seem the inescapable yet strikingly plausible result of a technological innovation. Many of these statements carry the further implication that the social consequences of our technical ingenuity are far-reaching, cumulative, mutually reinforcing, and irreversible.

An invention, once introduced into society, is thus depicted as taking on a life of its own. For example, the continuing improvement of the computer has followed a kind of internal logic (a logic embedded in its constituent material components and its design), so that each "generation" of enhanced computational sophistication has led, in a seemingly predetermined sequence, to the next. As the use of the computer spreads, more and more institutions have to reconfigure their operations to comport with the new capacities and constraints it creates. In the process, society as a whole becomes increasingly dependent on large, intricately interrelated technical systems. The whole network—a system of systems, or a megasystem—becomes the indispensable technological armature of the economy. Its continued functioning is a precondition for the reproduction of the entire social order.

Such a deterministic view of technology is a pervasive theme of the mass media nowadays. Take, for example, "The Machine That Changed the World," a 1993 documentary television series about the coming of the computer. The narrative structure is based on the stock before-and-after model, and the title neatly captures the idea—evidently appealing to large audiences—that advancing technology has a steadily growing, well-nigh irresistible power to determine the course of events. This version of the idea is what James calls "hard determinism."[2]

As the essays in this volume suggest, the idea of technological determinism takes several forms, which can be described as occupying places along a spectrum between "hard" and "soft" extremes. At the "hard" end of the spectrum, agency (the power to effect change) is imputed to technology itself, or to some of its intrinsic attributes; thus the advance of technology leads to a situation of inescapable necessity. In the hard determinists' vision of the future, we will have technologized our ways to the point where, for better or worse, our technologies permit few alternatives to their inherent dictates. To optimists, such a future is the outcome of many free choices and the realization of the dream of progress; to pessimists, it is a product of necessity's iron hand, and it points to a totalitarian nightmare.

Critics of "hard" determinism question the plausibility of imputing agency to "technology." After all, they argue, the word is merely a modern abstract noun for a certain category of the arts—what used to be called "the mechanic arts." There are hundreds of technologies, and few assertions about "technology" apply with equal validity to all of them. In spite of the existence of an engineering profession, technology is not an organized institution; it has no members or stated policies, nor does it initiate actions. How can we reasonably think of this abstract, disembodied, quasi-metaphysical entity, or of one of its artifactual stand-ins (e.g., the computer), as the initiator of actions capable of controlling human destiny? To note the reified character of this presumed agent is to recall that—until now, at least—no technology, no matter how ingenious and powerful, ever has initiated an action not preprogrammed by human beings.[3]

2. Ibid.

3. The proviso "until now" is included in deference to the claims often made nowadays on behalf of the imminent capacity of scientists and engineers, with the help of artificial intelligence, robotics, biogenetic technology, and artificial-life theory (or some combination thereof), to create a suprahumanly intelligent, self-directing, self-replicating agent, or "mind child," whose existence will in effect render obsolete the traditional boundaries between the mechanical and the organic, between art and nature. This claim may be seen, in fact, as the current terminus of one popular tradition of technological determinism.

At the other end of the spectrum, the "soft" determinists begin by reminding us that the history of technology is a history of human actions. To understand the origin of a particular kind of technological power, we must first learn about the actors. Who were they? What were their circumstances? This approach leads willy-nilly to the more exacting and productive questions in the historian's tool kit. Why was the innovation made by these people and not others? Why was it possible at this time and this place rather than another time or place? Who benefited, and who suffered? In lieu of a "hard" monocausal explanation for the genesis of the presumed determinative power of a technical innovation, these questions suggest the greater plausibility of a "soft," less specific, multivalent explanation. Instead of treating "technology" *per se* as the locus of historical agency, the soft determinists locate it in a far more various and complex social, economic, political, and cultural matrix.

The soft determinists' viewpoint may be illustrated by the way they might explain the growing credence given to the idea of technological determinism itself. An obvious historical starting point for this tendency is the marked acceleration in the rate of technical innovation that occurred, according to a broad current consensus of knowledgeable historians, in the West in the seventeenth and eighteenth centuries. But why did a propensity to innovate come to the fore at that time in the British Isles, in the North American colonies, and in Western Europe? Historians have proposed a great variety of well-documented, well-reasoned answers. Some focus on the particular efficacy of certain material, geographic, demographic, and socioeconomic preconditions: access to raw materials or markets; the existence of a mercantile capitalist economy; the operation of the profit motive; the accumulation of capital; the availability of a needy, teachable, exploitable labor force. Others attribute causal primacy to intellectual, cultural, or ideological factors: the extent of secular learning; the existence of a reservoir of entrepreneurial or financial skills; the presence of scientific rationalism, Christianity, the Protestant work ethic, or an artisanal ethos. Indeed, almost every identifiable attribute of early modern Western societies has been proposed as the putatively critical factor. Although it seems probable that the answer is to be found in some distinctive combination of these factors, the truth is that no one can say exactly what accounts for the special propensity to innovate that *initially* developed in the West in the early modern era.[4] Thus agency, as conceived by "soft" technological determinists, is deeply embedded in the larger social structure and culture—so deeply, indeed, as to divest technology of its presumed power as an independent agent initiating change.

And yet we need only look at the world of the 1990s to revivify the intuitively compelling idea that technological innovation is a major driving force of contemporary history, if not the primary driving force. Even if the critique of hard determinism is valid, it may only lead us to alter the status of technology to that of a second-order agent of history. Its power to effect change may be derived from certain specific socio-economic and cultural situations, but to say that is only to relocate the *origin* of that power. Once it has been developed, its determinative efficacy may then become sufficient to direct the course of events. In that case "technological determinism" has been redefined; it now refers to the human tendency to create the kind of society that invests technologies with enough power to drive history. If any particular form of human power now has an outstanding claim to that distinction, it probably is technological power. Indeed, one of our chief reasons for collecting these essays is our sense of the increasingly strong hold of that claim on the public imagination. People seem all too willing to believe that innovations in technology embody humanity's choice of its future. Whether that choice is an expression of freedom or an expression of necessity is the dilemma these essays are intended to elucidate.

Many of these essays were first delivered at a two-day workshop held at MIT in December 1989. In addition to the contributors to this volume, the participants included James Bartholomew, Nicholas Bloembergen, Alfred D. Chandler, Jr., I. Bernard Cohen, Jill K. Conway, Colleen Dunlavy, Gerald Holton, Robert Howard, Carl Kaysen, Kenneth Keniston, Philip Khoury, Bruce Mazlish, and William H. McNeill. Their contributions and interventions proved essential in helping us establish the topical and thematic outlines of the book.

4. The word "initially" requires special emphasis here because the recent development of technological sophistication in Japan, South Korea, Taiwan, and Singapore undermines any notion of an inherently or permanently distinctive affinity between the West and technological innovation.

As the project moved from the workshop to the compositional stage, we received advice and assistance from a number of people. Kenneth Keniston and Bronwyn Mellquist offered valuable commentary on several drafts of the introduction and the first essay. Pamela Laird, James H. Nottage, George O'Har, and Paul Vermouth provided much needed assistance with the selection and preparation of illustrations. An anonymous referee offered many helpful recommendations. We also wish to acknowledge the expert editorial assistance of Laurence Cohen and Paul Bethge of The MIT Press, whose close readings of the manuscript helped to improve our prose and clarify our ideas at many points. Finally, we wish to thank the Dibner Institute for the History of Science and Technology, particularly Executive Director Evelyn Simha and her staff, for sponsoring the workshop and helping to organize it. Our debt to the Dibner Institute is considerable, and the dedication on page v is meant to thank those who made it possible.

Does Technology Drive History?

Technological Determinism in American Culture

Merritt Roe Smith

In this essay, Merritt Roe Smith provides a brief history of technological determinism and shows how deeply such thought is embedded in American culture. He maintains that, as early as the 1780s, public servants like Tench Coxe began to attribute agency to the new mechanical technologies associated with the rise of the factory system. He reveals, moreover, that the technocratic spirit Coxe represented grew by leaps and bounds during the nineteenth century as the United States experienced rapid industrial expansion and gained status as a world power. Smith also shows how artists, advertisers, and professional historians contributed to the emergence of a widespread popular belief in technology as the driving force in society. He even detects elements of technological determinism in the writings of the scholars who became the most outspoken critics of modern technological society.

The belief in technology as a key governing force in society dates back at least to the early stages of the Industrial Revolution. Referred to as "technological determinism" by twentieth-century scholars, this belief affirms that changes in technology exert a greater influence on societies and their processes than any other factor. Indeed, one contemporary writer who refers to himself as a technocentrist maintains that "technology broadly conceived, along with its lesser sibling Science, is the central force in the modern world, more important to defining the patterns and problems of twentieth-century life than international conflict, national politics, the maldistribution of wealth, and differences of class and gender, because it is in some sense prior to all of these."[1] Within this genre of thought and expression one can discern two versions of technological determinism: a "soft view," which holds that technological change drives social change but at the same time responds discriminatingly to social pressures, and a "hard view," which perceives technological development as an autonomous force, completely independent of social constraints.[2]

The intellectual heritage of technological determinism can be traced to the enthusiasm and faith in technology as a liberating force expressed by leaders of the eighteenth-century Enlightenment. Within this tradition at least two streams of thought—one enthusiastic, the other critical—contributed to the formulation of the determinist position. Both viewpoints hold technology and science to be powerful agents of social change. This is noteworthy because deterministic thinking took root when people began to attribute agency to technology as a historical force. One sees such thought in the celebration of the new science by Voltaire, James Ferguson, and J. T. Desaguliers, in the memorable verses of Alexander Pope, in Diderot's *Encyclopedie,* in James Watt's ingenious feedback mechanisms, in the popular eighteenth-century metaphor of a clockwork universe, and

1. Memo from Wade Roush to the author, March 31, 1992 (in the author's possession).

2. Although he does not adopt the adjectives "soft" and "hard," Thomas J. Misa distinguishes between the various versions of technological determinism in "How Machines Make History, and How Historians (and Others) Help Them to Do So" (*Science, Technology, and Human Values* 13 (1988): 308–331) (see esp. p. 309). For a criticism and a denial of the distinction between hard and soft determinism, see Bruce Bimber's essay in this volume. Also see Alex Roland, "Theories and Models of Technological Change: Semantics and Substance," *Science, Technology, and Human Values* 17 (1992): 90–92.

even in the critical perspectives of such later essayists as Thomas Carlyle. Above all, deterministic thinking can be seen in the conception and popular acceptance during the eighteenth century of the idea of progress.[3]

While technological determinism initially sprouted in Europe, it found even more fertile ground in the newly independent United States—primarily because Americans were so taken with the idea of progress. Benjamin Franklin and Thomas Jefferson, foremost among the new nation's prophets of progress, were true believers in humankind's steady moral and material improvement. As avid proponents of the cause of liberty, they looked to the new mechanical technologies of the era as means of achieving the virtuous and prosperous republican society that they associated with the goals of the American Revolution. For them, progress meant the pursuit of technology and science in the interest of human betterment (intellectual, moral, spiritual) and material prosperity. Both men emphasized that prosperity meant little without betterment; a proper balance between them had to be maintained. Indeed, Jefferson's oft-repeated reservations about large-scale manufacturing reflected his concern about the fragility of liberty, power, and virtue in society and his sense of how easily a republic could be corrupted. If carried to extremes, Jefferson worried, the civilizing process of large-scale technology and industrialization might easily be corrupted and bring down the moral and political economy he and his contemporaries had worked so hard to erect. As much as Jefferson esteemed discovery and invention, he considered them *means* to achieving a larger social end. For his part, Benjamin Franklin refused to patent his inventions, for, as he put it, "we enjoy great advantages from the inventions of others, we should be glad of an opportunity to serve others by any invention of ours, and this we should do freely and generously." An exemplar of the Enlightenment, the author of Poor Richard's memorable aphorisms considered his inventions not a source of private wealth but a benefit for all members of society.[4]

3. For an introduction to the literature on technology and the idea of progress, see Merritt Roe Smith, "Technology, Industrialization, and the Idea of Progress in America," in *Responsible Science: The Impact of Technology on Society,* ed. K. B. Byrne (Harper & Row, 1986), and Leo Marx, "Does Improved Technology Mean Progress?" *Technology Review* (January 1987): 33–41, 71.

4. Smith, "Technology, Industrialization, and the Idea of Progress in Amer-

However, just when Franklin and Jefferson were espousing a new republican technology sensitive to human perfection, a more technocratic vision of progress was beginning to emerge. Evident in the speeches and writings of Alexander Hamilton and his associate at the U.S. Treasury Department, Tench Coxe, this new viewpoint openly attributed agency and value to the age's impressive mechanical technologies and began to project them as an independent force in society. Although Hamilton is remembered as the country's leading exponent of mechanized manufacturing during the 1790s, Coxe (1755–1824) was its most eloquent and persistent advocate. Like many of his contemporaries, Coxe believed that America's political independence hinged on the establishment of economic independence. Given the country's dependent economic status, he emphasized the need for machine-based manufactures as the prime solution to its political problems. Indeed, he told an audience of sympathetic listeners in the summer of 1787 that manufacturing under the factory system represented "the means of our POLITICAL SALVATION." "It will," he noted, "consume our native productions . . . it will improve our agriculture . . . it will accelerate the improvement of our internal navigation . . . it will lead us once more into the paths of virtue by restoring frugality and industry, those potent antidotes to the vices of mankind and will give us real independence by rescuing us from the tyranny of foreign fashions, and the destructive torrent of luxury."[5] Whereas Jefferson had emphasized technological development in the interest of the spiritual needs of individual citizens, Coxe shifted the emphasis away from human betterment and toward more impersonal societal ends, particularly the establishment of law and order in an unstable political economy. From the start,

ica," pp. 2–4; Marx, "Does Improved Technology Mean Progress?" pp. 35–37 (Franklin quote: p. 36).

5. Tench Coxe, *An Address to an Assembly of the Friends of American Manufactures, Convened for the Purpose of Establishing a Society for the Encouragement of Manufactures and the Useful Arts, Read in the University of Pennsylvania, on Thursday the 9th of August 1787*, reprinted in *The Philosophy of Manufactures: Early Debates Over Industrialization in the United States*, ed. M. B. Folsom and S. D. Lubar (MIT Press, 1982) (quotes from pp. 61–62). For perceptive treatments of Coxe's writings see John F. Kasson, *Civilizing the Machine* (Grossman, 1976), and Leo Marx, *The Machine in the Garden* (Oxford University Press, 1964), pp. 150–169.

technological determinism proved highly compatible with the search for political order.

As industrial capitalism gained a firmer grip on the American economy during the early decades of the nineteenth century, Coxe's technocratic perspective became increasingly dominant among other segments of the population.[6] While evidence of tension and discontent could be found among the new class of industrial workers and in the works of certain artists and intellectuals, by and large journalists, popular orators, and politicians hailed "the progress of the age," reassuring their audiences that technological innovation not only exemplified but actually guaranteed progress. The evidence for progress seemed incontrovertible. Decade by decade the pace of technological change quickened—railroads, steamships, machine tools, telegraphy, structures of iron and steel, electricity—and with each decade the popular enthusiasm for "men of progress" and for their inventions grew. Ralph Waldo Emerson, often a critic of the new mechanical age, exclaimed "Life seems made over new." "Are not our inventors," asked another enthusiastic writer, "absolutely ushering in the very dawn of the millennium?" It certainly seemed so to Horace Greeley, the editor of the *New York Tribune.* Upon visiting New York's Crystal Palace Exhibition in 1853, he pronounced: "We have universalized all the beautiful and glorious results of industry and skill. We have made them a common possession of the people. . . . We have democratized the means and appliances of a higher life." A writer in the prominent *North American Review* asserted that the benefits of machinery "are seen every where, felt every where, and must abide forever."[7]

What began as a trickle of enthusiasm for technology in the late eighteenth century became a rivulet by the Civil War and continued

6. Leo Marx provides an insightful discussion of the origins of the technocratic spirit in his contribution to this volume.

7. "Works and Days," in *Emerson's Works,* vol. 7, *Society and Solitude* (Houghton, Mifflin, 1870), p. 158; Douglas T. Miller, *The Birth of Modern America, 1820–1850* (Pegasus, 1970), p. 32; Horace Greeley, ed., *Art and Industry at the Crystal Palace* (Redfield, 1853), pp. 52–53; "Effects of Machinery," *North American Review* (Boston) 34 (January 1832): 226–227. For further documentation see Hugo A. Meier, "Technology and Democracy, 1800–1860," *Mississippi Valley Historical Review* 43 (1957): 618–640, and Russell B. Nye, *Society and Culture in America, 1830–1860* (Harper & Row, 1974), pp. 3–31, 258.

Figure 1
Christian Schussele's portrait "Men of Progress" (1863) celebrated the role of inventors in American society and sought to place them in the pantheon of national heroes. Among those portrayed are Samuel Colt (inventor of the revolving pistol bearing his name, third from left), Cyrus McCormick (fourth from left, standing beside a model of his reaper), Charles Goodyear (seventh from left, with a pair of rubber boots beneath his chair), and Elias Howe (extreme right, sitting behind his patented sewing machine). Samuel F. B. Morse (seventh from right) occupies the most prominent position in the portrait. He is seated at the table, demonstrating his telegraphic apparatus to the group, and has turned to converse with Robert Hoe (fifth from right), whose rotary printing press is represented by the drawing on the floor at Morse's feet. At upper left hangs a portrait of Benjamin Franklin, considered by many to be the symbol of American inventive genius and, appropriately in this instance, an important precursor of Morse in electrical experimentation. For further information, see Brooke Hindle and Steven Lubar, *Engines of Change* (Smithsonian Institution Press, 1986), pp. 75–77. Photograph courtesy of Dibner Institute for History of Science and Technology.

to surge thereafter. Between the 1860s and the early 1900s, many popular books, articles, paintings, and lithographs celebrating the new technology found their way into America's parlors and sitting rooms. A sampling of books from the period reveals such titles as *Eighty Years of Progress, Men of Progress, Triumphs and Wonders of the 19th Century, The Progressive Ages or Triumphs of Science, The Marvels of Modern Mechanism, Our Wonderful Progress, The Wonder Book of Knowledge*, and *Modern Wonder Workers.* The belief that in some fundamental sense technological developments determine the course of human events had become dogma by the end of the century. Writing in 1899, James P. Boyd confidently asserted that "the nineteenth century stands out in sublime and encouraging contrast with any that has preceded it." "As the legatee of all prior centuries," he continued, "it has enlarged and ennobled its bequest to an extent unparalleled in history." "Indeed," he concluded, "it may be said that along many of the lines of invention and progress which have most intimately affected the life and civilization of the world, the nineteenth century has achieved triumphs and accomplished wonders equal, if not superior, to all other centuries combined."[8]

8. James P. Boyd, *Triumphs and Wonders of the 19th Century* (C. W. Stanton, 1899), p. i. Also see, for example, *Eighty Years' Progress of the United States* (L. Stebbins, 1864); James H. Parton et al., *Sketches of Men of Progress* (New York and Hartford Publishing Co., 1870); Benson J. Lossing, *The American*

As much as late-nineteenth-century writers like Boyd admired the power, dynamism, and design of new technologies, they were even more taken with their symbolic qualities. Boyd and his collaborators described the latest developments in electricity, naval engineering, architecture, and chemistry in tones of awe and wonder. Coupled with advances in art and literature and with "the century's moral progress," such "glorious works" seemed "almost magical" to Boyd. He therefore entitled his book *Triumphs and Wonders of the 19th Century,* confident that the subjects he chose would "show the active forces, the upward and onward movements, and the grand results that have operated within, and triumphantly crowned, an era without parallel." At the heart of these triumphs stood the products of technological change.

Boyd's rhetoric may seem inflated to late-twentieth-century readers, but by the standards of the late nineteenth century it typifies the tendency of writers to view new technologies both as instruments of power and as triumphant symbols of human progress. Such writers expressed a seemingly unbounded enthusiasm for the machine age, so much so that one gets the impression that heavier and heavier doses of technology are being prescribed for the solution of societal ills. Inspired by their contacts with the great inventions of the age, writers and artists often purposely endowed steamboats, railway locomotives, machinery, and other inanimate objects with life-like qualities in order to cultivate emotions of wonderment, awe, magic, and, at times, even dread in their audiences. The discovery of what cultural historians would later call the "technological sublime" added yet another dimension to the growing popular belief in technology's power to shape the course of human history.[9]

Centenary: A History of the Progress of the Republic (Porter & Coates, 1876); Jerome B. Crabtree, *The Marvels of Modern Mechanism and Their Relation to Social Betterment* (King-Richardson, 1901); Trumbull White, ed., *Our Wonderful Progress: The World's Triumphant Knowledge and Works* (Hampden, 1902); H. L. Harvey, *The Progressive Ages or the Triumphs of Science* (J. A. Ruth, 1881); Alfred R. Wallace, *The Wonderful Century: Its Successes and Failures* (Dodd, Mead, 1898); Henry C. Hill, *The Wonder Book of Knowledge: The Marvels of Modern Industry and Invention* (John C. Winston, 1927); Waldemar Kaempffert, ed., *Modern Wonder Workers: A Popular History of American Invention* (Blue Ribbon Books, 1924); Malcolm Keir, *The Pageant of America: The Epic of Industry* (Yale University, 1926).

9. For an extended discussion of the technological sublime see Kasson, *Civilizing the Machine,* pp. 162–168.

Such themes are vividly illustrated in the popular arts of the era, particularly in the work of Currier and Ives of New York.[10] By all accounts the most prolific American lithographers of the nineteenth century. Currier and Ives issued well over 200 prints that depicted the centrality of steam power in shaping the American experience. In the next essay Michael Smith discusses one of their best-known works in this genre: the 1868 lithograph "Across the Continent: Westward the Course of Empire Takes Its Way," in which a steam-powered train of cars is depicted as the central force in the settlement of the American West and as the carrier of Anglo-Saxon values and civilization. As if to underscore the nation's destiny, the train is set on an endless trajectory, sweeping aside the remnants of Native American culture with its powerful puffs of dark smoke.

Other Currier and Ives lithographs suggest the power, elegance, beauty, and progressive character of machinery (see figures 2 and 3), but perhaps the most evocative tribute to technological progress and its ability to shape society is John Gast's 1872 oil painting entitled "Westward-ho" or "American Progress" (figure 4). In his commission, the patron-publicist George Crofutt instructed Gast to paint a "beautiful and charming female . . . floating westward through the air, bearing on her forehead the 'Star of Empire.'" "She has left the cities of the East far behind," Crofutt imagined, "and still her course is westward. In her right hand, she carries a book—common school—the emblem of education and the testimonial of our national enlightenment, while with the left hand she unfolds and stretches the slender wires of the telegraph, that are to flash intelligence throughout the land. . . . Fleeing from 'Progress' are the Indians, buffalo, wild horses, bears and other game, moving westward—ever westward. The Indians . . . turn their despairing faces toward the setting sun, as they flee from the presence of wondrous vision. The 'Star' is too much for them."[11]

10. For background information on Currier and Ives see Peter C. Marzio, *The Democratic Art: Chromolithography 1840–1900* (Godine, 1979), esp. pp. 59–63; John and Katherine Ebert, *Old American Prints for Collectors* (Scribner, 1974), pp. 121–168; Frederic A. Conningham, *Currier & Ives Prints: An Illustrated Check List* (1973; Crown, 1983), pp. v–xx, 4. For an insightful discussion of Currier and Ives' "Across the Continent" see Kasson, *Civilizing the Machine,* pp. 177–179.

11. Quoted from Christie's catalog of *Important American Paintings, Drawings, and Sculpture from the 19th and 20th Centuries* (1992), p. 48. Also see George Crofutt, *Crofutt's New Overland Tourist and Pacific Coast Guide* (Overland, 1878)

Figure 2
Currier and Ives, "American Express Train" (1864). Photograph from The Old Print Shop, New York.

As art goes, "American Progress" is not a work of great distinction. But as a popular allegory that amalgamates the idea of America's Manifest Destiny with an old republican symbol (the goddess Liberty, now identified as Progress) and associates progress with technological change (represented by the telegraph lines, the railroads, the steamships, the cable bridge, and the urban landscape in the background), it is a remarkable achievement. The painting clearly conveys the dominant culture's attitude toward nature, Native Americans, and, more generally, linear change and improvement through science and technology. As the art historian Joshua Taylor observes, the maiden

and J. V. Fifer, *American Progress: The Growth of the Transport, Tourist and Information Industries in the Nineteenth-Century West, Seen Through the Life and Times of George A. Crofutt, Pioneer and Publicist of the Transcontinental Age* (Globe Pequot Press, 1988).

Figure 3
Currier and Ives, "The Progress of the Century" (1876). Photograph from Library of Congress.

Progress's movement "brings with it all the blessings of technology and settled life in an orderly and rapid passage across space and time." "It is a grand and majestic progress," Taylor continues, "and everybody is happy with it except that snarling bear and those wretched Indians in the lower lefthand corner." The visibility—even dominance—of various forms of technology in Gast's painting (and in many other artworks of the period) suggests that the idea of progress was becoming increasingly technocratic.[12]

If Gast's maiden Progress still bore some trappings of early republican culture in 1872, she became a full-blown technocrat by the turn

12. Joshua C. Taylor, *America as Art* (Smithsonian Institution Press, 1976), pp. 145–146. With reference to the maiden Liberty as symbol of America's

Figure 4
John Gast, "American Progress" (oil on canvas, 1872). Photograph courtesy of Gene Autry Western Heritage Museum, Los Angeles.

of the century. The frontispiece of James Boyd's 1899 volume *Triumphs and Wonders of the 19th Century* presents an allegory of progress in which Liberty—now called Industry—is surrounded by the artifacts of technological progress (figure 5). What is striking about this print is the extent to which political and artistic symbols are dominated by technological ones. Indeed, the only explicit classical republican symbols—the Corinthian columns—are off to one side and are identified as representing architecture rather than politics. The goddess Electricity is elevated over and outshines her companions Music, Literature, and Art. In the eyes of the artist, technology had assumed a place of dominance in American culture.

Early in the twentieth century, as artists and writers continued to tout the machine as the *primum mobile* of society, technocratic thinking received an enormous boost from the burgeoning field of professional advertising. Characterized by historians as the distinctive institution of American technological culture, advertising became the instrument by which big business, in need of ever-expanding markets for its mass-produced products, imprinted instrumental values—and, with them, the ethos of mass consumption—on the populace. Advertising agencies, in short, not only sold the products of industrial capitalism but also prompted a way of thinking about industrial technology. Using the psychological concepts of association and suggestion, neatly packaged in colorful and briefly worded appeals that excited mental images, advertisers encouraged people to believe that technology, broadly construed, shaped society rather than the other way around. As more and more psychology-based advertisements reached the public through print, radio, and eventually television, technology became idolized as the force that could fix the economy and deliver on the legendary promise of American life. Such technocratic pitches constituted a form of technological determinism that embedded itself deeply in popular culture.[13]

youthful freedom and virtue, Taylor notes: "Although the mighty eagle was a comforting symbol of unity and power, it did not occupy the field alone. Liberty references rarely use the eagle as standing for the new country. It was the family of goddesslike personifications of social virtues that remained the most persuasive images of what the new United States signified."

13. On the origins of modern advertising see Michal McMahon, "An American Courtship: Psychologists and Advertising Theory in the Progressive Era," *American Studies* 13 (fall 1972): 5–18; David M. Potter, *People of Plenty: Economic Abundance and the American Character* (University of Chicago Press,

Figure 5
"Triumphs and Wonders of the Nineteenth Century" (1899), unsigned frontispiece from James P. Boyd's book of the same title. According to the author: "This picture explains and is symbolic of the most progressive one hundred years in history. In the center stands the beautiful female figure typifying Industry. To the right are the goddesses of Music, Electricity, Literature and Art. Navigation is noted in the anchor and chain leaning against the capstan; the Railroad, in the rails and cross-ties; Machinery, in the cog-wheels, steam governor, etc.; Labor, in the brawny smiths at the anvil; Pottery, in the oranamental vase; Architecture, in the magnificent Roman columns; Science, in the figure with quill in hand. In the back of picture are suggestions of the progress and development of our wonderful navy. Above all hovers the angel of Fame ready to crown victorious Genius and Labor with the laurel wreaths of Success." Photograph from Graphic Arts Service, MIT.

With the tools of the new applied psychology, the advertisers of the early twentieth century quickly mastered the idea of the technological fix. Their aim was to excite consumers' needs by associating them with intangible desires. They therefore developed advertisements that appealed to such thoughts and feelings as efficiency, elegance, family affection, freedom, modernity, patriotism, sexuality, status, and youth. In April 1905, for example, the makers of Rubens Infant Shirts and Cream of Wheat placed ads in the *Ladies Home Journal* that pressed the issue of child health and appealed to the nurturing instincts of mothers to feed and clothe their children properly. With the advent of electrical appliances, Simplex marketed its power ironers by aiming ads at the "intelligent" housekeeper who courageously sought to "meet the overwhelming demands upon her time" by making ironing easier. The advertisement shown in figure 8, and countless others like it, assured readers that the products of the new technology not only saved time but also made users "happier, healthier, and more cheerful when the work is finished." Technology had now become the *cause* of human well-being.[14]

1954), esp. chapter 8 ("The Institution of Abundance: Advertising"); Roland M. Marchand, *Advertising the American Dream* (University of California Press, 1985).

14. See Ruth S. Cowan, "The 'Industrial Revolution' in the Home: Household Technology and Social Change in the 20th Century," *Technology and Culture* 17 (January 1976): 1–42; Cowan, *More Work for Mother* (Basic Books, 1983), pp. 137–138 and 187–188 and figure 8.

Figure 6
Advertisement for Rubens Infant Shirt, *Ladies' Home Journal*, April 1905. Photograph from Graphic Arts Service, MIT.

Figure 7
Advertisement for Cream of Wheat, *Ladies Home Journal*, April 1905.

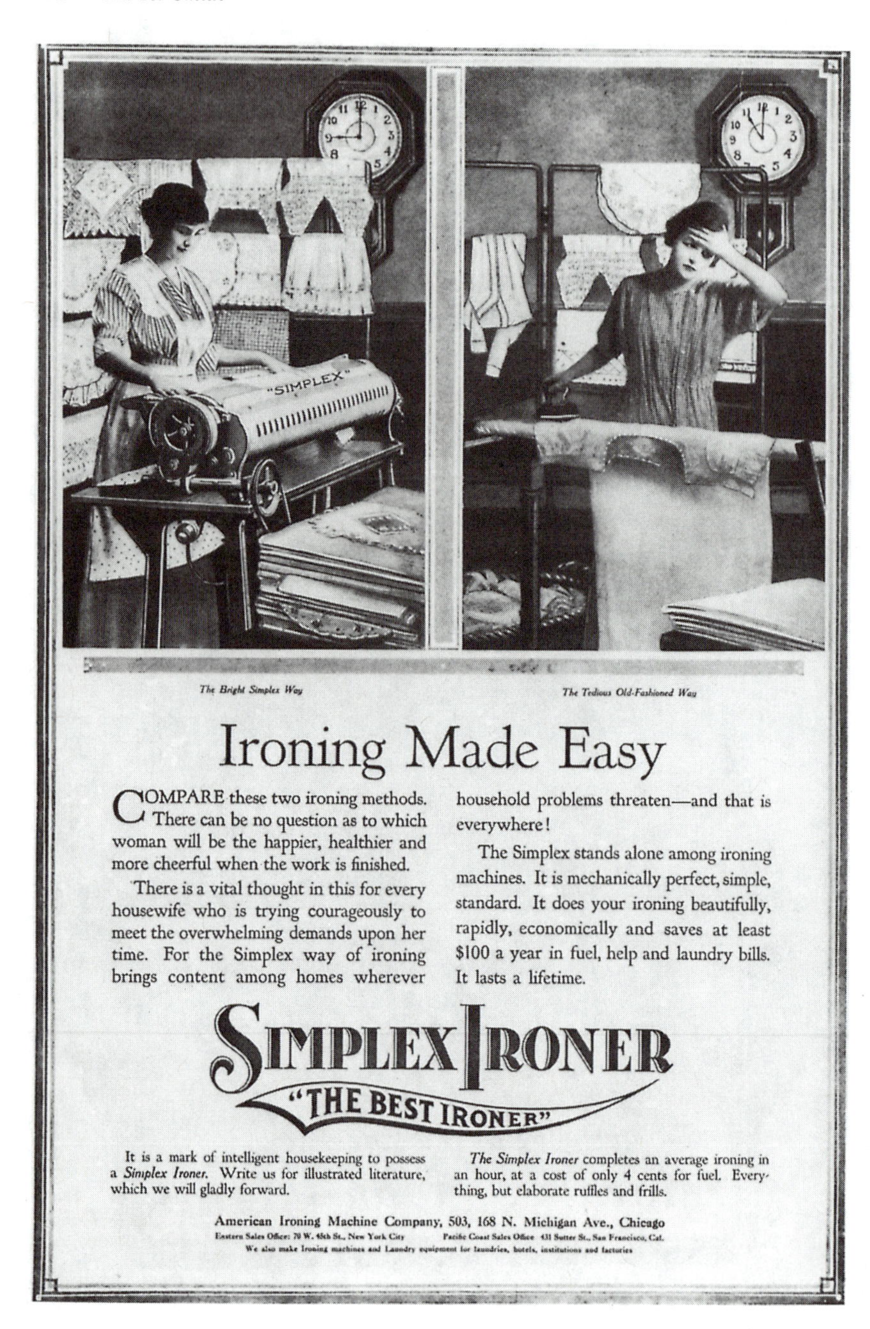

Figure 8
Advertisement for Simplex Ironer, *Ladies Home Journal*, April 1920. Photograph from Cleveland Public Library.

Nowhere were the benefits of technology more evident than in automobile advertisements. In addition to identifying their products with the usual psychological appeals, automakers let pass no opportunity to point out the latest technological features of their products and the advantages they would bring to buyers. A 1940 Chrysler ad in the *Ladies Home Journal* hailed fluid (automatic) drive as "the next best thing to a chauffeur" (figure 9). About to get into the "beautiful" vehicle was a woman smartly attired in a business suit, white gloves, and a fur stole. Standing on the other side of the car, as if in a mirage or a dream, was a handsome liveried chauffeur giving a salute. "Be modern," the ad urged, "buy Chrysler!" Being modern meant many things—beauty, elegance, status, smartness—but most of all it meant, in this instance, ease of operation made possible by the latest advances in technology. Other advertisements, (such as that shown in figure 10, which is from 1954) carried similar messages.

Many advertisements went beyond offering products as technological fixes, however, From the early 1900s onward, advertising agencies sold the public on the idea that the latest advances in technology brought not only immediate personal gains but also social progress. Ads associated products and their makers with the progress of the age.[15] Toward the end of World War II, for example, the Bendix Aviation Corporation ran advertisements that assured readers that "tomorrow's cars are forming now in the crucible of war" and that "Bendix—through long association with the automotive industry and broad experience in many industrial and scientific fields—is particularly well equipped to speed this progress" (figure 11). The Crane

15. For a good example of the sort of celebratory publications that appeared during the bicentennial of the American Revolution, see *Iron Age* 217 (April 12, 1976). Similar publications exist for various world's fairs and for other national commemorations. Helpful entry points to this literature are the following: Merle Curti, "America at the World Fairs, 1851–1893," *American Historical Review* 55 (1950): 833–856; Robert C. Post, ed., *1876: A Centennial Exhibition* (Smithsonian Institution, 1976); Robert W. Rydell, *All the World's a Fair: Visions of Empire at American International Expositions, 1876–1916* (University of Chicago Press, 1984); Robert W. Rydell, *World of Fairs: The Century of Progress Expositions* (University of Chicago Press, 1993); Michael L. Smith, "Death of the Future: Depictions of Technology at the 1964 World's Fair," paper delivered at annual meeting of Society for the History of Technology, Cleveland, 1990.

Figure 9
Advertisement for Chrysler automobile, *Ladies Home Journal*, April 1940. Photograph from Cleveland Public Library.

Figure 10
Advertisement for General Motors' Autronic-Eye, *Saturday Evening Post*, February 20, 1954. Photograph from Graphic Arts Service, MIT.

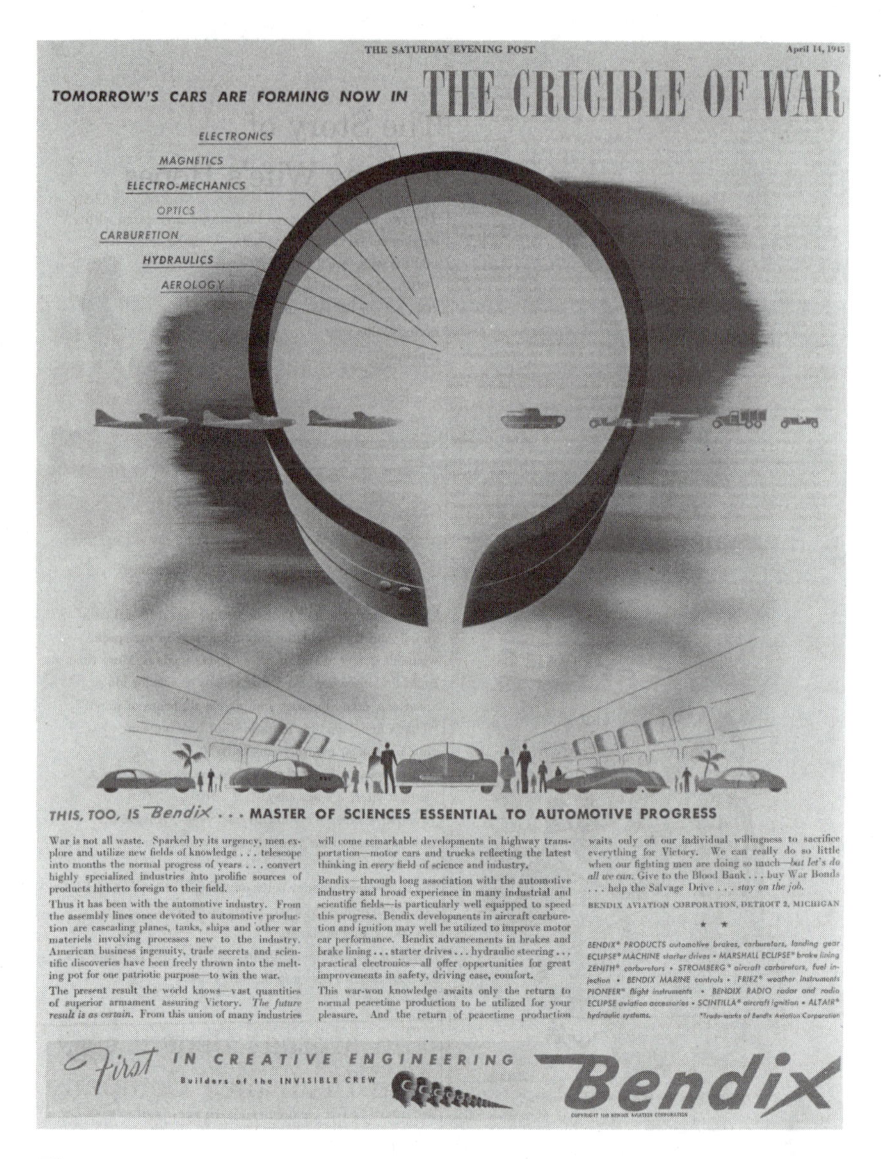

Figure 11

Bendix Aviation's "crucible of war" advertisement, *Saturday Evening Post*, April 14, 1945. Photograph from Graphic Arts Service, MIT.

Company celebrated the approach of its hundredth anniversary in 1952 by pointing to the many contributions it had made to "industrial progress" in a two-page ad in *Popular Mechanics* (figure 12). To the "baby boomers" who grew up during the 1950s and 1960s, such slogans as "Progress is our most important product" and "Better living through chemistry" became part and parcel of the American Dream. Journalists and scholars began to refer to the United States as "the technological society," and most Americans embraced a creed that—in contrast with Jefferson's concept of progress—glorified the march of invention and the material progress of the age. Technology and science not only became the great panacea for everyday problems; they also stood for values at the core of American life.[16]

Many tributaries flowed into the stream of thought known as technological determinism, and no brief introductory essay can do justice to them all. As central as political thought, art, literature, journalism, and advertising are to the formation of the technocratic perspective, other sources deserve at least brief mention. More should be said about the technological utopian literature that flourished in the late nineteenth and early twentieth centuries and made technological advance central to the perfection of society. More also should be said of the Technocracy movement, which enjoyed brief popularity during the 1930s and then receded in the public eye, only to resurface periodically after the Second World War. Such ideas and movements, coupled with a growing historical literature that celebrated the nation's industrial development and ignored its darker aspects, helped to consolidate the hold that technocratic thought exercised on the popular imagination.[17]

16. A well-known contemporary example is EPCOT Center, to which John Staudenmaier refers in his essay below. Also see Michael L. Smith, "Back to the Future: EPCOT, Camelot, and the History of Technology," in *New Perspectives on Technology and American Culture,* ed. B. Sinclair (American Philosophical Society, 1986).

17. For helpful treatments of technological utopianism see Howard P. Segal, *Technological Utopianism in American Culture* (University of Chicago Press, 1985); Kasson, *Civilizing the Machine,* chapter 5. On the technocracy movement, see Segal, pp. 120–128; William E. Akin, *Technocracy and the American Dream: The Technocratic Movement, 1900–1941* (University of California Press, 1977); Henry Elsner, Jr., *The Technocrats: Prophets of Automation* (Syracuse University Press, 1967). For a sampling of popular historical works that embody deterministic thinking, see Waldemar Kaempffert, ed., *A Popular*

Contributing to Industrial Progress since 1855
On July Fourth, 1855, a new industry was born. In his foundry in Chicago, R. T. Crane poured the molten metal to make the first Crane casting. America was on the threshold of a mighty era of progress—an era that would show dramatic advance in the development of power—in transportation—in industrial production. The valves and fittings manufactured by Crane Co. have played a vital part in the building of industrial America. Below are a few of the many fields that Crane has served for nearly a century.
IN OIL PRODUCTION—Coal oil for lamps—this was the petroleum industry in 1855. Crane piping has served in the mighty advances made in the production, as well as the refining, of oil and in the processing of the many petroleum by-products.
IN RAIL TRANSPORTATION—It's a far cry from the chugging little wood burner of 1855 to the slick streamlined diesel of today. On locomotives and in passenger cars, in shops and roundhouses, in fact, wherever piping is used, Crane Valves and Fittings are serving the nation's railroads.

Figure 12
Crane Company's centennnial advertisement, *Popular Mechanics*, January 1952. Photograph from Graphic Arts Service, MIT.

Although technological determinism and its analogue, the technocratic idea of progress, became dominant perspectives in the United States and in other industrial societies, they nonetheless encountered criticism. Early on, an adversary culture arose to challenge technocratic thought, both for its instrumental values and for its conception of history. One sees such criticism in Thomas Jefferson's warnings about the introduction of the factory system; in Thomas Carlyle's attack on the "mechanical philosophy"; in the writings of Ralph Waldo Emerson, Nathaniel Hawthorne, Herman Melville, and Henry David Thoreau; in the landscape art of Thomas Cole, George Innes, Andrew Melrose, and Thomas Rossiter; and in such dystopian novels as Mark Twain's *Connecticut Yankee in King Arthur's Court* and Ignatius Donnelly's *Caesar's Column.*[18]

The critics worried that Americans, in their headlong rush to mechanize and rationalize production, were sacrificing moral progress for material power, thus abandoning a concern that was central to thinkers of Jefferson's generation. With rapid industrial growth, critics feared, the country was drifting away from its revolutionary republican moorings toward a more secular and materialistic frame of belief. Emerson, who had once embraced invention and the "mechanics arts" as expressions of "Young America's" genius and vitality, grew increasingly restive. "What have these arts done for the character, for the worth of mankind?" he asked an audience in 1857. "Are men better?" The answer seemed clear to Emerson: "'Tis too plain that with the material power the moral progress has not kept pace. It appears that we have not made a judicious investment. Works and days were offered us, and we took works." Thoreau agreed. "Men have become tools of their tools," he wrote in *Walden.* In their awe and enthusiasm for new inventions and the money-getting system that spawned them, more and more people sought "but improved

History of American Invention, two vols. (Scribner, 1924); Malcolm Kier, *The Epic of Industry* (Yale University Press, 1926); Roger Burlingame, *March of the Iron Men* (Scribner, 1938) and *Engines of Democracy* (Scribner, 1940); John W. Oliver, *History of American Technology* (Ronald Press, 1956); Daniel J. Boorstin, *The Americans: The Democratic Experience* (Random House, 1973).

18. Leo Marx, *The Machine in the Garden* (Oxford University Press, 1964), esp. pp. 117–144, 170–179, 227–319; Susan Danly and Leo Marx, eds., *The Railroad in American Art* (MIT Press, 1988); Kasson, *Civilizing the Machine,* chapter 5; Smith, "Industry, Technology, and the Idea of Progress," pp. 7–9.

means to an unimproved end."[19] In Thoreau's view, they had made a Faustian bargain.

Another important critic of the new technology was Henry Adams. In his famous autobiography, Adams wrote movingly about the fundamental changes that separated modern America from its premodern past, beginning with the observation that "the old universe was thrown into the ash heap and a new one created [in 1844 with] the opening of the Boston and Albany Railroad; the appearance of the first Cunard steamers in the bay; and the [reception of] telegraphic messages which carried . . . the news that Henry Clay and James K. Polk were nominated for the Presidency."[20]

The capstone of Adams' discontinuous personal and symbolic experience, however, was the trip he made, in the company of Dr. Samuel Langley of the Smithsonian Institution, to the Paris Exposition of 1900, where he witnessed the tremendous invisible power generated by electric dynamos. Awed by the experience, Adams reported that he "began to feel the forty-foot dynamos as a moral force, much as the early Christians felt the Cross." Moreover, he sensed that the dynamo had replaced the cross as the primary force in civilization. Indeed, he found himself praying to it! For Adams, the contrast between the dynamo and the cross symbolized an enormous shift of faith away from the great principles of Christianity, and religion generally, toward those of science and utility. The former stood for love; the latter for power.

For Adams, as for Emerson and Thoreau, the contrast between these two "kingdoms of force" spoke volumes about what had been lost through industrialization. Nearly 40 years earlier, in 1862, Adams had observed to his brother Charles that "man has mounted science and is now run away with." "I firmly believe," he continued, "that before many centuries more, science will be the master of man. The engines he will have invented will be beyond his strength to control." By 1900, the reality of that statement seemed even more apparent. Upon witnessing the automobile (which "had become a nightmare at a hundred kilometres an hour") and learning about the recent dis-

19. "Works and Days," in *Emerson's Works,* vol. 7, *Society and Solitude* (Houghton, Mifflin, 1870), p. 166; Henry David Thoreau, *Walden* (1854; Modern Library, 1981), pp. 33 and 46. Also see Marx, *Machine in the Garden,* pp. 229–265.

20. *The Education of Henry Adams* (1918; Modern Library, 1931), p. 5.

covery of radioactivity (which "denied its God—or, what was to Langley, the same thing, denied the truths of his Science"), Adams "found himself lying in the Gallery of Machines at the Great Exposition of 1900, his historical neck broken by the sudden eruption of forces totally new."[21]

Compared with the highly personal, moral, and nostalgic tone of nineteenth-century commentators, twentieth-century writers seem more detached and impersonal (but no less committed) in their criticisms of the technocratic perspective. Matters of faith and tradition mean less to them than questions of politics and power. Three individuals merit special attention here because their work has become so central to the emergence of a vigorous adversary culture. They are Lewis Mumford, Jacques Ellul, and Langdon Winner.[22]

Of the three, Mumford produced the largest and most varied body of work. While his early research focused mainly on the history of architecture, literature, and urban planning, in 1934 he published *Technics and Civilization,* a study of tremendous breadth and insight that placed technology in a larger cultural context and argued, in effect, that culture preceded technics in human evolution. Among other things, Mumford pointed to Western religious monasticism and warfare as forces that sparked and spread new technologies and ultimately led to the rise of modern industrial capitalism, with its emphasis on rationality, regimentation, and order. But like his Enlightenment forebears, Mumford worried about the threat technological progress posed to social and spiritual progress. In the end he consoled himself with the hope that the new science-based technologies of the twentieth century would reconcile the differences between the machine and the human spirit through the emergence of an "organic mechanism" that would help humankind realize its highest potential.[23]

21. Ibid., pp. 380–382; Adams to Charles Francis Adams, April 11, 1862, reprinted in *The Letters of Henry Adams,* ed. J. C. Levenson et al. (Harvard University Press, 1982), 1: 290.

22. While I have chosen to focus my remarks on Mumford, Ellul, and Winner, they certainly are not the only critics of technocratic culture. Notable contributions have been made by Herbert Marcuse, Martin Heidegger, Rene Dubos, Paul Goodman, Murray Bookchin, Kurt Vonnegut, David F. Noble, and David Dickson.

23. See Mumford, *Technics and Civilization* (1934; Harcourt Brace and World,

By the early 1950s, Mumford's earlier optimism had dissolved into a bitter and strident condemnation of the technocratic perspective. "Like a drunken locomotive engineer on a streamlined train, plunging through the darkness at a hundred miles an hour," he observed in *Art and Technics,* "we have been going past the danger signals without realizing that our speed, which springs from our mechanical facility, only increases our danger and will make more fatal the crash." In Mumford's view, "our overmechanized culture" had fallen prey to the most dangerous betrayal of all—the "Myth of the Machine"—and, despite the defeat of fascism in World War II, was rapidly moving toward a "final totalitarian structure." In the competition for world markets, industrial societies pressed hard to develop technological capacities that would give them an edge and, in the process, made the machine rather than the human condition the norm against which all else was measured. "Our age is passing from the primeval state of man . . . to a radically different condition," Mumford wrote dejectedly in the prologue to his last major work. The first state was "marked by his invention of tools and weapons for the purpose of achieving mastery over the forces of nature." Under the new dispensation "he will have not only conquered nature, but detached himself as far as possible from the organic habitat." "With this new 'megatechnics,'" Mumford lamented, "the dominant minority will create a uniform, all-enveloping, super-planetary structure, designed for automatic operation. Instead of functioning actively as an autonomous personality, man will become a passive, purposeless, machine-conditioned animal whose proper functions, as technicians now interpret man's role, will either be fed into the machine or strictly limited and controlled for the benefit of de-personalized, collective organizations."[24]

For Mumford, such depersonalized organizations constituted the 'megamachine,' an all-inclusive yet "invisible" entity that included the technical and scientific apparatus as well as the bureaucratic hierarchy

1963). Also see Mumford, "Democratic and Authoritarian Technics," *Technology and Culture* 5 (winter 1964): 1–8; Arthur Molella, "Inventing the History of Invention," *American Heritage of Invention and Technology* 4 (spring–summer 1988): 26–28.

24. Lewis Mumford, *Art and Technics* (Columbia University Press, 1952), pp. 11–12; Mumford, *The Myth of the Machine,* vol. 1, *Technics and Human Development* (Harcourt Brace Jovanovich, 1966); pp. 3, 189, 200.

that organized and controlled it. Ever conscious of the gender implications of his argument, Mumford identified the megamachine's bureaucracy as consisting of "a group of men, capable of transmitting and executing a command, with the ritualistic punctilio of a priest, the mindless obedience of a soldier." Given the power of the megamachine and the invidious threat it posed, mankind's only hope was to seize the moment and take back what had been abdicated. "For those of us who have thrown off the myth of the machine," Mumford concluded, "the next move is ours. . . . Each one of us, as long as life stirs in him, may play a part in extricating himself from the power system by asserting his primacy as a person in quiet acts of mental or physical withdrawal—in gestures of non-conformity, in abstentions, restrictions, inhibitions, which will liberate him from the domination of the pentagon of power."[25]

The ominous message conveyed by Lewis Mumford became even more ominous in the hands of Jacques Ellul, who asserted in his 1954 book *La Technique* (subsequently translated and published in English as *The Technological Society*) that "technique has become autonomous," that "no human activity escapes this technical imperative,"[26] and that the human race has been swept up in the all-encompassing power of technique. "Technique," Ellul wrote, "has become the new and specific *milieu* in which man is required to exist. . . . It is artificial, autonomous, self-determining, and independent of all human intervention."[27]

As numerous commentators have noted, the word "technique," as Ellul used it, refers to a good deal more than machines. Technique includes machines and other technological devices, to be sure. But it also encompasses organizational methods, managerial practices, and, most important of all, a mode or manner of thinking that is inherently mechanistic. "Technique integrates the machine into society," Ellul wrote. "It constructs the kind of world the machine needs and introduces order. . . . It clarifies, arranges, and rationalizes; it does in the

25. Mumford, *Technics and Human Development,* pp. 188–194, 200; Mumford, *The Myth of the Machine,* vol. 2, *The Pentagon of Power* (Harcourt Brace Jovanovich, 1964), pp. 420, 430, 433–35.

26. Jacques Ellul, *The Technological Society* (1964; Vintage, 1967), pp. 6, 14, 21.

27. Ellul, "The Technological Order," *Technology and Culture* 3 (fall 1962): 10.

domain of the abstract what the machine did in the domain of labor. It is efficient and brings efficiency to everything. . . . In fact, technique is nothing more than *means* and the *ensemble of means*. . . . Our civilization is first and foremost a civilization of means; in the reality of modern life, the means, it would seem, are more important than the ends. Any other assessment of the situation is mere idealism."[28]

For Ellul, the technological society he described, if not a *fait accompli,* was very close to being so. While he identified and described the new milieu, he did not applaud it. In fact, in the foreword to the revised American edition of *The Technological Society* Ellul described his work as a diagnosis of a disease. Yet when it came to prescribing a treatment or solution to the problem, Ellul wrote that he "[did] not yet know" what was to be done. His purpose was "to arouse the reader to an awareness of technological necessity and what it means"—to sound "a call to the sleeper to awake."[29]

That the sleeper would awake in time was, in Ellul's view, possible but not likely. Large if not insuperable obstacles faced those who wished to liberate themselves from the totalitarian embrace of Technique. Ellul concluded that "as long as man worships Technique, there is as good as no chance at all that he will ever succeed in mastering it," and that "if the *individual* cannot attain personal liberty with respect to technical objects, there is no chance that he will be able to respond to the general problem of Technique." In the end, Ellul's message, if not fatalistic, is very pessimistic.[30]

Not so Langdon Winner's. In his first book, *Autonomous Technology,* Winner hewed close to the path Ellul had blazed. Echoing Ellul, he defined "the idea of autonomous technology" as "the belief that somehow technology has gotten out of control and follows its own course, independent of human direction." However, Winner's thought differs from Ellul's in several important respects. Whereas Ellul depicted technique as a highly rational, all-embracing governing force, Winner detected a much more erratic and volatile phenomenon. Hence the subtitle of his book: *Technics-out-of-Control as a Theme in Political Thought.*[31]

28. Ellul, *The Technological Society,* pp. 5, 19.

29. Ibid., pp. xxxi, xxxiii.

30. Ellul, "The Technological Order," pp. 25–28.

31. Langdon Winner, *Autonomous Technology: Technics-out-of-Control as a Theme in Political Thought* (MIT Press, 1977), p. 13.

"We do not *use* technologies as much as *live* them," Winner argued. "Highly developed complex technologies are tools without handles or, at least, with handles of extremely remote access." The reason for this is that "men release powerful changes into the world with cavalier disregard for consequences; that they begin to 'use' apparatus, technique, and organization with no attention to the ways in which these 'tools' unexpectedly rearrange their lives; that they willingly submit the governance of their affairs to the expertise of others . . . [and] begin to participate without second thought in megatechnical systems far beyond their comprehension or control; . . . that they stand idly by while vast technical systems reverse the reasonable relationship between means and ends." According to Winner, people have "come to accept an overwhelmingly passive response to everything technological." Human "somnambulism," rather than any inherent technological imperative, has allowed large technological systems "to legislate the conditions of human existence."[32]

In today's world, technology legislates. But, in Winner's view, it need not be that way. In perhaps his most important insight, he maintained that "technology is itself a political phenomenon" and, as such, is potentially subject to political constraints. In *Autonomous Technology* Winner advanced a series of proposals based on a "decentralized democratic politics" and aimed at dismantling problematic technological systems "with the expressed aim of studying their interconnections and their relationships to human need."[33] His most recent book, *The Whale and the Reactor,* articulates this position in greater detail and places it in a deeper historical context. A central point is that technological systems, with their inherent political qualities, are not value neutral. Indeed, they invariably favor the interests of some over the interests of others. Winner consequently maintains that societies, if they are to be equitable and effective, must understand precisely what sorts of implications new technologies may carry with them before they are introduced. He concludes:

The important task becomes . . . not that of studying the 'effects' and 'impacts' of technical change, but one of evaluating the material and social infrastructures specific technologies create for our life's activity. We should try to imagine and seek to build technical regimes compatible with freedom, social

32. Ibid., pp. 202, 314, 324.

33. Ibid., pp. 323–335.

justice, and other key political ends. . . . Faced with any proposal for a new technological system, citizens or their representatives would examine the social contract implied by building the system in a particular form. They would ask, How well do the proposed conditions match our best sense of who we are and what we want this society to be? Who gains and who loses power in the proposed change? Are the conditions produced by the change compatible with equality, social justice, and the common good? To nurture this process would require building institutions in which the claims of technical expertise and those of a democratic citizenry would regularly meet face to face. Here the crucial deliberations would take place, revealing the substance of each person's arguments and interests. The heretofore concealed importance of technological choices would become a matter for explicit study and debate.[34]

The fundamental problem, in Winner's view, is not technological determinism but rather "the often painful ironies of technical choice." In contrast with the essentially elitist formulations of Ellul, Winner proposes a democratic political philosophy by which to plan and implement new technologies. Moreover, his emphasis on reestablishing the priority of ends over means has a distinctly Jeffersonian-Enlightenment quality. In this respect he extends and deepens Lewis Mumford's call to replace authoritarian with democratic technics. But, unlike Mumford, he advocates activist interventions in the political process rather than "quiet acts" of individual protest or other forms of symbolic behavior. Though Winner's argument has been criticized as too theoretical and lacking in specificity, his emphasis on moral and political principles goes a long way toward formulating programs for action. To be sure, Winner is deeply concerned about the threats existing technological systems pose to the natural environment and the human condition. But he is confident that the search for reasonable limits to growth will succeed. Essentially, his critique of the technological society ends on a hopeful note.[35]

34. Langdon Winner, *The Whale and the Reactor: A Search for Limits in an Age of High Technology* (University of Chicago Press, 1986), pp. 55–56.

35. Ibid.; Winner, "Upon Opening the Black Box and Finding it Empty: Social Constructivism and the Philosophy of Technology," in *The Technology of Discovery and the Discovery of Technology: Proceedings of the Sixth International Conference of the Society for Philosophy and Technology,* ed. J. C. Pitt and E. Lugo (Society for Philosophy and Technology, 1991), pp. 503–519. Also see Mary O'Connell's review of *The Whale and the Reactor* (*Bulletin of the Atomic Scientists* (March 1987): 57–58).

The writings of Mumford, Ellul, and Winner have an ironic twist: in speaking out against the pervasive power of technological systems and the serious threats they pose both to humanity and nature, these critics have endowed modern technics with a degree of agency and influence that often goes beyond even what its most enthusiastic advocates claim. Thus, to the extent that they place technology at the forefront of social and cultural change, Mumford, Ellul, and Winner are technological determinists. Yet even here certain distinctions must be made between Ellul's avowedly determinist position and the more nuanced and carefully delineated stances of Mumford and Winner. Such contrasts are important because in recent years historians of technology have been critical of writers who invoke "the demands" of technology as a causal force in history.[36] Because Mumford and Winner underscore issues of technological choice and depict technologies as social products as well as social forces, they have generally avoided the criticism that has greeted Ellul. In short, their emphasis on the contextual settings of technological change comes very close to the current governing paradigm among historians of technology.[37]

The essays in this volume are less concerned with the sort of technological determinism expressed by Ellul, Mumford, and Winner than with technocratic thought as a culturally embedded attitude. The latter encompasses the former, to be sure, but it differs from the former in that it often uncritically equates technological change with progress. As we have seen, this is definitely not the case with

36. The historians Judith A. McGaw, David F. Noble, and John M. Staudenmaier rank among the most outspoken critics of technological determinism. See, for example, McGaw's *Most Wonderful Machine: Mechanization and Social Change in Berkshire Paper Making, 1801–1885* (Princeton University Press, 1987), pp. 4–7; Noble's *Forces of Production: A Social History of Automation* (Knopf, 1984), pp. xi–xiii, 324–326; and Staudenmaier's "Perils of Progress Talk: Some Historical Considerations," in *Science, Technology, and Social Progress,* ed. S. Cutcliffe and S. Goldman (Lehigh University Press, 1989), pp. 268–298, and "Science and Technology: Who Gets a Say?" in *Technological Development and Science in the Industrial Age: New Perspectives on the Science-Technology Relationship,* ed. M. Becker and P. Kroes (Kluwer, 1992), pp. 205–230.

37. On the significance of contextualism in the history of technology, see John M. Staudenmaier, *Technology's Storytellers* (MIT Press, 1985), esp. chapters 4 and 5; Thomas P. Hughes, "From Deterministic Dynamos to Seamless-Web Systems," in *Engineering as a Social Enterprise,* ed. H. E. Sladovich (National Academy Press, 1991), pp. 7–25.

Ellul, Mumford, and Winner. Because the technocratic perspective springs from so many sources and is so pervasive in society, it is very difficult to pin down. Like the Chinese *budaoweng* doll that bounces back up each time someone knocks it down, the technocratic perspective is extremely resilient. Hence the thrust of this essay, which seeks to set the problem, identify the antagonists, and supply a context in which to read and appreciate the contributions that follow. Understanding this context helps to explain why the theme of technological determinism continues to attract, and frustrate, scholars. Who among us would deny that it is easy to be drawn into technology-driven explanations of cultural and historical processes? The frustration we experience with technological determinism and technocratic thinking is as much with ourselves as it is with the larger societies in which we live. As moths to a flame, we find ourselves continually attracted to its alluring but dangerous glow.

Recourse of Empire: Landscapes of Progress in Technological America

Michael L. Smith

Public commentators often remark that Americans had an abiding love affair with the machine. In this essay Michael Smith asks how they came "to attach so much of their national and personal identity to technology." At some level, he contends, the belief that technology brings inevitable social progress is a myth, the product of "a curious cultural and political fetishism." Yet at another level he indicates that the idea of technology as progress gained widespread currency "precisely because it could be depicted as carving an uncontested, inevitable path." According to Smith, media representations of technology are crucial to understanding both how public opinion about technology forms and how opinions change. He illustrates this process by comparing two "landscapes of progress": a Currier & Ives lithograph from 1868 and a Leydenfrost drawing that appeared in a 1952 issue of Popular Mechanics. *He maintains that Currier & Ives emphasize technological progress as an open-ended source of economic growth and cultural integration whereas Leydenfrost stresses innovation, novelty, and power in an unimaginable future ("What will they think of next?"). Indeed, Smith notes, Leydenfrost's "frenzy of detail nearly obscures the fact that we don't know the shape of things to come." Yet, ironically, at the very time* Popular Mechanics *celebrated the "march" of science and invention, new uncertainties about the nuclear age clouded the future.*

One of the social effects of technological change is that it prompts cultures to wonder where they are heading. In the United States, generations of leaders and pundits have mistaken technology for the answer, rather than the question. The artifacts of technological innovation—electric lights, automobiles, airplanes, personal computers—have come to signify progress, as well as the ever-receding goals toward which we are said to be progressing. In countless ads and speeches, twentieth-century Americans have been asked to visualize the future as a succession of unimaginable new machines and products.

How did technology come to represent so many aspects of national and personal identity in American culture? And how has technology evolved under the weight of such expectations? To pose such questions is to venture into the murky bog of technological determinism, where machines seem to have lives of their own.

"Technological determinism" is a curious phrase. The gist of it is heartbreaking in its simplicity: the belief that social progress is driven by technological innovation, which in turn follows an "inevitable" course. I imagine that it would be captivating to see a world so pure in design, so subversive in effect. I say subversive because the notion of inevitable, technology-driven progress upends the Second Law of Thermodynamics, conjuring up a non-entropic universe in which dust never settles, things cohere and grow supple with time, and everything is always new and improved, forever young. There is a joyful, renegade quality to this view of the world—as if twentieth-century Americans had become technological Bonnies and Clydes, rushing headlong against the wind, confident that their car would always be the fastest and that they would never be caught. (And isn't that, after all, what advertising has always promised?)

We have to suspect, though, that Bonnie and Clyde knew all along that it was only a matter of time before the contradictions they had set in motion (defying the powers that constricted them, yet hoping to find serenity in the eye of the hurricane) would tear them apart. Perhaps such self-knowledge is buried within our culture as well. At some level, postwar Americans may have suspected for some time that faith in technology-as-progress can serve as a substitute for a more genuine participatory democracy. Without a clear sense of alternative routes, however, sheer momentum can steer the chase.

We scholars of technology and culture lament the stubborn tenacity of technological determinism, but we rarely try to identify the needs it identifies and attempts to address. On the face of it, our brief against this variety of superstition resembles the academy's response to creationism: How can something so demonstrably wrong-headed continue to sway adherents?

The comparison is instructive. The periodic resurgences of creationist sentiment over the past century have not really been about evolution, or the fossil record, or the depths of geological time. Those who insist that their children (and everyone else's) be taught that the Biblical creation myth is (or at least might be) literally true may simply be trying to reclaim authority in a world where every category of human enterprise is controlled by experts. What was true during the Scopes trial of the 1920s remains applicable to the textbook debates of the 1980s and the 1990s: issues of class and autonomy underlie competing definitions of science and authority. (Ironically, latter-day creationists have adopted the language, credentials, and tactics of their opponents, dredging up their own "scientists" and "experts" to argue that the world really is only 6000 years old.)

My intention is not to defend creationism or technological determinism, but rather to suggest that their adherents share a social terrain and a vocabulary of representation—not only with one another but also with those who denounce them. Those of us who regard either biblical or technological fundamentalism as superstition have had to shed some of the self-assurance with which we once dismissed these misconceptions as less preferable than our own. We have had to embrace and expand upon Thomas Kuhn's now-hoary insight that science spins its own creation myths (and, we might add, fosters its own superstitions)—one major difference being that the pronouncements of science tend to be revised more frequently than those of religion.

About technological determinism we could also argue that the issue is not really technology at all but rather a curious cultural and political fetishism whereby artifacts stand in for technology, and technology in turn signifies national progress. Perhaps, in industrialized societies, technologies are visible primarily by means of the trappings with which each culture dresses them. To understand technology as lived experience, we need to acquire a comparative view of how different cultures perceive, define, and meet technological challenges and opportunities. Historians of the United States can contribute to this task by tracing the evolution of the gospel of technological progress and by seeking clues to the ways in which Americans have adopted, exploited, or resisted it.

Some aspects of technology periodically split into conflicting ideological camps. A good example is the new American nation's debate, two centuries ago, over systems of weights and measures. The proponents of the Continental (metric) system emphasized the appeal of adopting a system alien to the hated British oppressors; but the argument that won the day was that the English system was more democratic, in that it was more accessible to mechanics and merchants who could not afford weighing and measuring equipment. Goods could be divided into half-pounds or quarts by a careful eye; not so tenths. Inches and yards roughly corresponded to the dimensions of knuckles and arms—but centimeters? (And while ideological importance changes over time, ingrained habits are difficult to alter. In the twentieth century, Congress "corrected" the weights and measures issue by adopting the metric system—and then completely ignored its enforcement.)

In the corridors of power, however, these exceptions have proved the rule. By and large, generations of the nation's political and corporate leaders have spoken in the sweeping cadences of technological determinism because of its irresistible power to cast them as trailblazers on the uncontested, inevitable path of progress. And as technology became more complex, and the decisions behind its development receded from view, the public increasingly had to rely on *images* of technology for its understanding of both technology and progress. The history of those images is very different from the history of the technology itself.

The techniques and media for representations of technology have themselves undergone sweeping changes; the Chatauquas and mechanics' fairs and lithographs of the nineteenth century have given way to mass-market advertising, globally broadcast lunar expeditions, personal computer networks, and the universalization of video. Two images of American landscapes—one from just after the Civil War, and one from early in the post-World War II era—demonstrate some of the ways in which the depiction of change itself has changed.

In 1868, Currier & Ives issued one of their most famous lithographs: "Across the Continent: Westward the Course of Empire Takes Its Way" (figure 1). The simple landscape it depicted was a grid on which Victorian-era Americans could locate the emblems and the direction of their nation's progress. Across a flat, green swath of

Figure 1

open plains, a railroad crosses the landscape from the lower right to the vanishing point in the upper left. To the left of the rail line, in the foreground, settlers are busily constructing a town. A cluster of homes and a public school are visible, and the few remaining trees are being cleared. Beyond the tiny village, Conestoga wagons can be seen heading west into the empty expanse.

On the other side of the tracks, Indians on horseback ponder the scene before them. A thick plume of smoke from the iron horse is about to engulf them. In the distance, another Indian canoes across a tree-lined waterway, and a range of mountains scrape the clouds.

The segregation of natives and settlers on the two sides of the rail line provided an alternative to depicting the actual relation between these two groups—not because Currier & Ives were reluctant to portray the extermination of Native Americans, but because such a scene would have distracted from the theme of the lithograph: the smooth, unchallenged inevitability of the train's progress across the terrain. Beneath its calm and uncontested surface, this vision of westering-as-progress signified conquest over the land, its natural elements, and its inhabitants.

In 1868, when this lithograph appeared, railroad crews from opposite ends of the continent were laboring toward each other in the final frantic months before the completion of the first transcontinental railroad. The image combines frontier mythology with state-of-the-art technology, capturing them at a moment when the geographic frontier had not yet disappeared and when technology could still be depicted primarily as an extension of existing social patterns and values. That moment, however, was destined to give way to very different depictions of technology and society. The railroad, which so neatly symbolized the progress of empire in 1868, soon became contested terrain. From the Great Upheaval of 1877 to the Pullman strike of 1894, railroad workers struggled to stop the erosion of their autonomy that accompanied the new industrial order.

Even many of the settlers on the prairie decided that the railroad, with its monopoly on shipping rates and its control over vast tracts of land, was a curse, not a blessing. More than anything else, hatred of the railroad, and of the social organization of technology and capital it represented, fueled the great nineteenth-century grassroots movement that came to be called Populism. If we could fast-forward the lithograph, the plume of smoke might be blowing toward both sides of the tracks, this time threatening to engulf not just the Indians but also the settlers, and even the operators of the train.

Not only the train but the landscape it traversed was poised to change. The 1890 census declared that the American frontier was gone; twentieth-century Americans would have to find a substitute for westward migration to signify the nation's progress. In the absence of a geographical frontier, mainstream American culture placed new emphasis on an alternate iconographic terrain: the technological frontier, where the speeding train would appear, not as the new conveyer of progress, but as progress itself.

In January 1952, *Popular Mechanics* published its Golden Anniversary Issue, celebrating the scientific and technological breakthroughs of the previous 50 years and looking ahead to the next 50. MIT president Karl Taylor Compton contributed a feature article titled "Science on the March," for which *Popular Mechanics* commissioned a two-page "symbolic drawing by A. Leydenfrost, the noted illustrator"—a landscape of progress depicting technological changes since the magazine's founding in 1902 (figure 2).[1]

A mid-twentieth-century counterpart to "Westward the Course of Empire," Leydenfrost's "Science on the March" illustration also depicts a terrain bisected by a railroad, with a steam locomotive making its way across the landscape. However, the 1952 rendition of American progress differs from its 1868 predecessor in several respects.

The Currier & Ives lithograph locates the future in the distance, in an unarticulated terrain that melts into the vanishing point. As viewers, we are asked to imagine ourselves moving with the train toward the dimly glimpsed future. The vanishing point is nearly blank, waiting to be "filled in" by the train, the settlers, and the viewer's eye, all moving toward closure. The *Popular Mechanics* landscape, by contrast, locates the *past* at the vanishing point (although everything before the turn of the century is beyond our scope of vision). The locomotive and the other artifacts of "progress" rush toward the viewer, exploding outward toward some unseen and unimaginable future located beyond the confines of the image. It is a dynamic image of progress, but we can't see where it is headed;

1. Karl T. Compton, "Science on the March," *Popular Mechanics* 97 (January 1952): 120.

1942
1932
1922
1912
1902

Figure 2

the frenzy of detail nearly obscures the fact that we don't know the shape of things to come.

Where "Westward the Course of Empire" implied conquest by introducing a single train onto a broad natural horizon, "Science on the March" covers every inch of the terrain and the sky with artifacts of industrial technology; the only remnant of the natural world is a small cage of laboratory rabbits. Even the train is nearly lost in the proliferation of industrial production.

The vanishing point of the *Popular Mechanics* image is dominated by a radiantly backlit nineteenth-century barn and silo, from which everything else in the picture appears to emanate. The barn might easily have been lifted from the Currier & Ives landscape; but an explanatory chart identifies this structure as "GE's first research laboratory located in a barn." Thus a frontier artifact has been transformed into a starting point for modern science and technology. The new pioneers are dressed in lab coats and armed with slide rules.

The landscape of "Westward the Course of Empire" is organized around the steam locomotive, which represents the settlers' most complex and modern technological tool for conquering space. But the physical presence of the train does not overpower the image; the expanse of plains *contextualizes* it and defines technology in terms of the frontier. In the "Science on the March" illustration, that formula is inverted: the barn, signifying the frontier, is redefined in terms of science and technology.

Parallel to Leydenfrost's locomotive, a streamlined diesel train charges toward us. And alongside the diesel, automobiles roll into the foreground, clearly overtaking the train as the dominant vehicle and moving through time as well as space; the early horseless carriages in the background "evolve" into 1952 models, leading up to a hyper-streamlined "car of the future" in the foreground.

As in the 1868 lithograph, the rail lines and highway divide the landscape into two areas. To the left, research equipment spills toward us from the barn. In the middle distance we see two "early scientists at work" (the only people who appear anywhere in the panorama). In the foreground are stylized representations of research in chemistry, nuclear physics, and electronics.

Across the tracks looms a city. No residences are depicted, however—only commerce and government. As the cityscape moves toward the present, brick and granite give way to ever-taller skyscrapers. "Manmade bolts of lightning" dance above an immense, windowless research laboratory, flanked by a cyclotron. Towering streamlined silos are identified as a "factory for synthetic products."

Even the sky is an assembly line of ever-more-advanced aircraft streaming toward the viewer. Like the automobiles, they evolve from pre-World War I biplanes to a two-engine monoplane, a dirigible, a "flying wing," a four-engine plane, jets, and a helicopter. At the front of this procession, the future is represented by a rocket ship emerging from a gigantic solar-system model of the atom; beyond them, ring-shaped and spiral nebulae stream across the sky.

In "The Moose on the Wall," the essayist Edward Hoagland tells of a young boy who enters a bank in northern Vermont, sees the stuffed head of a moose mounted on the wall, and asks to go into the adjoining room "so that he could see the rest of the animal."[2] Both of the landscapes that we have been examining ask us to picture "the rest of the animal"—the past or the future. Perhaps they could serve that purpose for each other. If we could mount the two images back to back on opposite sides of the same wall, with the two vanishing points neatly aligned, we could then board the train in the foreground of the Currier & Ives lithograph and ride it through to the other side, disembarking at the foreground of the Leydenfrost illustration. In the process, we would witness a continuum of strategies for representation of progress in industrializing American culture. (The most interesting part of the journey, the point at which technology overcomes westering as the embodiment of progress, would be obscured by a tunnel inside the wall, between the two vanishing points.)

Contemporaries of Currier & Ives would not have been puzzled by the blank expanse spread before them; the "future" would simply extend familiar settlement and city-building enterprises westward. Readers of the January 1952 issue of *Popular Mechanics,* however, would have had greater difficulty imagining the "future" that the drawing elects not to portray. Instead, they—and we—must extrapolate from past marvels.

These images, of course, are only two among many. Their different strategies for representing progress, however, point to some of the problems underlying the emerging iconographic role of technology in American culture. One reason "Science on the March" invokes the

2. Edward Hoagland, "The Moose on the Wall," in *The Courage of Turtles* (Random House, 1971).

future by depicting the past is that by 1952 images of technology's social identity overwhelmingly stressed innovation rather than integration. The ongoing social effect of systems of techniques (the everyday lives of people in an advanced industrial culture) was far less apparent than the promise of novelty; thus "Science on the March" presents an inexhaustible conveyor belt of ever newer and more advanced products.

This depiction of progress was consistent with the need for more and bigger markets, and its promise of consumer involvement provided a substitute for participatory decision-making. But when novelty is what is being fetishized, how does one visualize it? The persuasive force of "What will they think of next?" as a national creed rests in the assurance of unimaginable wonders just beyond the horizon. In such a landscape, anything that exists is already an artifact of past wonders, and therefore ineligible as a signifier for the unimaginable. One common solution is to invoke the future by analogy with the past; the audience is asked to imagine a future as different from now as now is from, say, 1902. Leydenfrost's illustration succinctly captures this approach.

But other strategies for representing the future were available, even if social objectives (or the absence thereof) were obscured from view. One alternative approach to representing the unimaginable—popular in corporate advertising and at world's fairs—was to transpose familiar futurist landscapes to unfamiliar settings: deserts, the ocean floor, outer space. This strategy lent an aura of novelty to old ideas and technologies. (The Future World exhibit at Disney World's EPCOT Center are elaborations of this vision.[3])

The "Science on the March" drawing, however, asks the viewer to complete the outward trajectory of its parade of progress. The only hints of the future are a stylized atom, a rocket, and an automobile—precisely the areas of scientific and technological innovation that were most prominently featured in popular media in the 1950s. (*Popular Mechanics,* for example, devoted a special column to each of them.) And just as automobile advertising promised to enhance the consumer's personal attributes, publicity for the "peaceful atom" and the space program promised to add stature to the nation's image.[4]

3. See Roland Marchand and Michael L. Smith, "Corporate Science on Display," in *Science and Social Reform in Industrial America,* ed. R. Walters (Johns Hopkins University Press, forthcoming).

4. Michael L. Smith, "Selling the Moon: The U.S. Manned Space Program

In 1952 a mere reference to nuclear and space technologies connoted impending wonders, but disparities between these images and their social possibilities were already difficult to avoid. Even when depicted as a clean, abstract celestial form, the atom reminded mid-twentieth-century Americans that the problem of visualizing the unimaginable had acquired a troubling underside. The new uncertainties of the nuclear age permeated visions of technology with a tacit acknowledgment that some new unveilings might be less welcome than new-model cars.[5] Is that rocket at the edge of the *Popular Mechanics* illustration a nuclear-powered Conestoga, extending human exploration into new frontiers? Or is it an ICBM? Viewers could imagine two undepicted futures beyond the frame of the image: the cornucopia of wonders might continue to expand, or it might dissolve into nuclear war. If "Westward the Course of Empire" sidestepped the genocide and appropriation imbedded in its image of progress, perhaps "Science on the March" conveys, without acknowledging, the contradictions inherent in Cold War visions of technology.

Less apparent in 1952 were the subtler threats awaiting this triumvirate of technologies of the future. In December 1953, President Eisenhower told the United Nations that a new "atoms for peace" program would transform the "greatest of destructive forces" into "a great boon for the benefit of all mankind." The following September, he waved a "radioactive wand" that broke ground for the nation's first commercial nuclear power plant at Shippingport, Pennsylvania.[6] Thirty-five years later, the core of the decommissioned Shippingport reactor was shipped to the Hanford Nuclear Reservation in Washington for burial. (At the same time, Hanford—a key facility for nuclear weapons production since World War II—was being con-

and the Triumph of Commodity Scientism," in *The Culture of Consumption: Critical Essays in American History, 1880–1980,* ed. R. W. Fox and T. J. Jackson Lears (Pantheon, 1983).

5. Paul Boyer, *By the Bomb's Early Light: American Thought and Culture at the Dawn of the Atomic Age* (Pantheon, 1985); Spencer Weart, *Nuclear Fear: A History of Images* (Harvard University Press, 1988); Michael L. Smith, "Advertising the Atom," in *Government and Environmental Politics: Essays on Historical Developments Since World War Two,* ed. M. J. Lacey (Wilson Center Press, 1989).

6. *New York Times,* December 9, 1953, p. 2; "A Wand Wave, A New Era," *Life* 37 (September 20, 1954): 141.

verted into a massive environmental cleanup operation to minimize the severe health and safety threats resulting from nearly a half-century of inadequate safety standards and careless waste disposal, at an estimated cost of between 30 billion and 50 billion dollars.) Like the decommissioning, the construction and operation of the small Shippingport plant had been heavily subsidized by the federal government. Its operating costs were 10 times those of a coal-fired plant; its "showcase" decommissioning distracted attention from the fact that the rest of the nation's nuclear plants (many of them much larger) would require dismantling without such lavish federal assistance.[7]

While the Shippingport reactor was in operation, major controversies arose over cooling systems, radioactive waste disposal, and worker and community safety; the "peaceful atom" became as controversial as its military counterpart. The accidents at Three Mile Island in 1979 and at Chernobyl in 1986, along with previously unreleased information about earlier accidents around the world, reversed long-standing public support for nuclear power in the United States.

The trajectory of space exploration also proved uneven. The Soviet Union's launch of Sputnik I in October 1957 had sent Congress scurrying to create NASA and to land Americans on the moon. In the early years of the space program, huge budget allocations for space exploration were justified in terms compatible with the "Science on the March" conveyor belt of unimaginable wonders. Testifying before the House of Representatives' Select Committee on Astronautics and Space Exploration in the spring of 1958, rocket engineer Wernher von Braun observed that the exploration of outer space, like "scientific progress" in general, was propelled by the need to envision the unknown: "People are just curious. . . . What follows in the wake of their discoveries is something for the next generation to worry about."[8]

7. Seth Shulman, "When a Nuclear Reactor Dies, $98 Million Is a Cheap Funeral," *Smithsonian* 20 (October 1989): 56–69; Keith Schneider, "Military Has New Strategic Goal in Cleanup of Vast Toxic Waste," *New York Times*, August 5, 1991, pp. A1, C3.

8. *Hearings before the Select Committee on Astronautics and Space Exploration, House of Representatives, on H.R. 11881, 85th Congress, 2nd Session, April–May 1958* (Government Printing Office, 1958), p. 38.

What the next generation discovered was that, in the absence of clear post-Apollo goals, the U.S. space program suffered from the militarization of its agenda, from the curtailment of poorly planned major projects (Skylab and the Space Station), and from a series of malfunctions and errors—most notably the 1986 *Challenger* disaster.[9]

Not even the automobile, sacred emblem of American technology and successor to the train of progress, escaped reassessment in the years just beyond the frame of "Science on the March." In 1952, few could have doubted that Leydenfrost's endless succession of ever-bigger models would continue indefinitely. Not far down that road, however, the car faced safety challenges, air-pollution standards, energy crises, fuel-efficiency requirements, and a growing awareness in congested cities that the automobile could no longer substitute for effective mass transit. And while American automakers spent millions lobbying to block new government standards, foreign competitors exceeded those standards and won control of the market.[10]

In spite of the proliferation of challenges to unexamined optimism, Americans have been reluctant to discard their vision of technology-as-progress. That impulse should be respected and understood. In the absence of greater access to decision-making, citizens have been confined to the role of consumers, lacking the shared capacity to view gradations of social possibilities for technology. Too often, our vocabulary has been limited to oversimplified pro- vs. anti-technology sentiments. Even moderate calls for more circumspect technological assessment have been routinely met by a defensiveness that reveals how clouded our vision of technology can be by cultural preconceptions. (In 1970, Atomic Energy Commissioner Theos J. Thompson characterized questions regarding the safety of nuclear technology as a direct challenge to "the American philosophy of life."[11])

Yet we seem to be slowly emerging from what Langdon Winner has termed "technological somnambulism"—awakening to the real-

9. Paul B. Stares, *The Militarization of Space: U.S. Policy, 1945–1984* (Cornell University Press, 1985).

10. Marcia D. Lowe, "Rethinking Urban Transport," in *State of the World 1991*, ed. L. Brown et al. (Norton, 1991); Martin V. Melosi, *Coping with Abundance: Energy and Environment in Industrial America* (Knopf, 1985).

11. Theos J. Thompson, "Improving the Quality of Life—Can Ploughshare Help?" *Symposium on Engineering with Nuclear Explosives, January 14–15, 1970*, vol. 1 (Atomic Energy Commission), p. 104.

ization that as a national culture we are not Daniel Boone, or Bonnie and Clyde, or even the crew of the Starship *Enterprise*. We have more nearly resembled a latter-day Walter Mitty, jarred from our day-dreaming to discover that the landscapes we drive through, the fuel we consume, and the air we pollute do not get left behind as we drift on to yet another "final frontier."[12] Our elusive destination may turn out to be nearer than we thought. In the 1960s, Marshall McLuhan celebrated the "global village"; we have since discovered that all of us who inhibit that village are "downwinders," consuming the blessings and the curses of technology with every breath. Our new frontiers are not geographic.

What would an update of "Science on the March" look like in 2002? Historians are reluctant to describe the future, but I am happy to propose a utopian fantasy to help fill the void that will surely follow the demise of technological determinism: a terrain characterized, not by the headlong commuter-rush toward distant horizons of progress, but by a homegrown version of *The Day the Earth Stood Still.* For a brief period, everything stops. Everyone gets out of the car of the future, walks about, talks with everyone else about what needs to be done. They begin to devise their own maps for roads not taken. They agree that social expertise requires collective effort and should no longer be confused with technical expertise. They redefine national security, dismantling Stealth bombers and SDI laser satellite systems and converting them into prenatal care, universal health coverage, safe energy, clean air and water, reliable mass transit, and the elimination of hunger. They elect to replace pesticides with Integrated Pest Management techniques. They begin a conversion project to remove the great dams from Western rivers and to minimize the environmental effects of manufacturing. Defense lobbyists, sensing that their jobs are now superfluous, choose to become historians, and set about recording the stories unfolding all around them, in the national interest.

12. Langdon Winner, *The Whale and the Reactor: A Search for Limits in an Age of High Technology* (University of Chicago Press, 1986), p. 10.

Do Machines Make History?

Robert L. Heilbroner

Among the scholars represented in this anthology, economic historian Robert Heilbroner comes closest to embracing technological determinism. But he does so with carefully stated qualifications. Heilbroner's now-classic 1967 essay seeks to specify the extent to which technology determines "the nature of the socioeconomic order." He does so by showing that technological change follows a roughly ordered sequence of development and that it "imposes certain social and political characteristics upon the society in which it is found." Yet, rather than take a strong stand for technological determinism, Heilbroner opts for a "soft" version, because he sees a complex historical scenario in which technology, while acting on society, also reflects the influence of socioeconomic forces on its development. Ultimately he views technology as a strong "mediating factor" rather than as the *determining influence on history—a point he reiterates and expands upon in the retrospective essay that follows this one.*

"Do Machines Make History" first appeared in Technology and Culture *(8 (July 1967): 335–345).*

"The hand-mill gives you society with the feudal lord; the steam-mill, society with the industrial capitalist."
—*Marx,* The Poverty of Philosophy

That machines make history in some sense—that the level of technology has a direct bearing on the human drama—is of course obvious. That they do not make all of history, however that word be defined, is equally clear. The challenge, then, is to see if one can say something systematic about the matter, to see whether one can order the problem so that it becomes intellectually manageable.

To do so calls at the very beginning for a careful specification of our task. There are a number of important ways in which machines make history that will not concern us here. For example, one can study the impact of technology on the *political* course of history, evidenced most strikingly by the central role played by the technology of war. Or one can study the effect of machines on the *social* attitudes that underlie historical evolution: one thinks of the effect of radio or television on political behavior. Or one can study technology as one of the factors shaping the changeful content of life from one epoch to another: when we speak of "life" in the Middle Ages or today we define an existence much of whose texture and substance is intimately connected with the prevailing technological order.

None of these problems will form the focus of this essay. Instead, I propose to examine the impact of technology on history in another area—an area defined by the famous quotation from Marx that appears above. The question we are interested in, then, concerns the effect of technology in determining the nature of the *socioeconomic order.* In its simplest terms the question is: Did medieval technology bring about feudalism? Is industrial technology the necessary and sufficient condition for capitalism? Or, by extension, will the technology of the computer and the atom constitute the ineluctable cause of a new social order?

Even in this restricted sense, our inquiry promises to be broad and sprawling. Hence, I shall not try to attack it head-on, but to examine it in two stages:

1. If we make the assumption that the hand-mill does "give" us feudalism and the steam-mill capitalism, this places technological change in the position of a prime mover of social history. Can we then explain the "laws of motion" of technology itself? Or, to put the

question less grandly, can we explain why technology evolves in the sequence it does?

2. Again, taking the Marxian paradigm at face value, exactly what do we mean when we assert that the hand-mill "gives us" society with the feudal lord? Precisely how does the mode of production affect the superstructure of social relationships?

These questions will enable us to test the empirical content—or at least to see if there *is* an empirical content—in the idea of technological determinism. I do not think it will come as a surprise if I announce now that we will find *some* content, and a great deal of missing evidence, in our investigation. What will remain then will be to see if we can place the salvageable elements of the theory in historical perspective—to see, in a word, if we can explain technological determinism historically as well as explain history by technological determinism.

I

We begin with a very difficult question hardly rendered easier by the fact that there exist, to the best of my knowledge, no empirical studies on which to base our speculations. It is the question of whether there is a fixed sequence to technological development and therefore a necessitous path over which technologically developing societies must travel.

I believe there is such a sequence—that the steam-mill follows the hand-mill not by chance but because it is the next "stage" in a technical conquest of nature that follows one and only one grand avenue of advance. To put it differently, I believe that it is impossible to proceed to the age of the steam-mill until one has passed through the age of the hand-mill, and that in turn one cannot move to the age of the hydroelectric plant before one has mastered the steam-mill, nor to the nuclear power age until one has lived through that of electricity.

Before I attempt to justify so sweeping an assertion, let me make a few reservations. To begin with, I am fully conscious that not all societies are interested in developing a technology of production or in channeling to it the same quota of social energy. I am very much aware of the different pressures that different societies exert on the direction in which technology unfolds. Lastly, I am not unmindful of the difference between the discovery of a given machine and its application as a technology—for example, the invention of a steam engine (the aeolipile) by Hero of Alexandria long before its incorporation into a steam-mill. All these problems, to which we will return in our last section, refer, however, to the way in which technology makes its peace with the social, political, and economic institutions of the society in which it appears. They do not directly affect the contention that there exists a determinate sequence of productive technology for those societies that are interested in originating and applying such a technology.

What evidence do we have for such a view? I would put forward three suggestive pieces of evidence:

The Simultaneity of Invention

The phenomenon of simultaneous discovery is well known.[1] From our view, it argues that the process of discovery takes place along a well-defined frontier of knowledge rather than in grab-bag fashion. Admittedly, the concept of "simultaneity" is impressionistic,[2] but the related phenomenon of technological "clustering" again suggests that technical evolution follows a sequential and determinate rather than random course.[3]

The Absence of Technological Leaps

All inventions and innovations, by definition, represent an advance of the art beyond existing base lines. Yet, most advances, particularly in retrospect, appear essentially incremental, evolutionary. If nature

1. See Robert K. Merton, "Singletons and Multiples in Scientific Discovery: A Chapter in the Sociology of Science," *Proceedings of the American Philosophical Society* 105 (October 1961): 470–486.

2. See John Jewkes, David Sawers, and Richard Stillerman, *The Sources of Invention* (New York, 1960 [paperback edition]), p. 227, for a skeptical view.

3. "One can count 21 basically different means of flying, at least eight basic methods of geophysical prospecting; four ways to make uranium explosive; . . . 20 or 30 ways to control birth. . . . If each of these separate inventions were autonomous, i.e., without cause, how could one account for their arriving in these functional groups?" S. C. Gilfillan, "Social Implications of Technological Advance," *Current Sociology* 1 (1952): 197. See also Jacob Schmookler, "Economic Sources of Inventive Activity," *Journal of Economic History* (March 1962): 1–20; Richard Nelson, "The Economics of Invention: A Survey of the Literature," *Journal of Business* 32 (April 1959): 101–119.

makes no sudden leaps, neither, it would appear, does technology. To make my point by exaggeration, we do not find experiments in electricity in the year 1500, or attempts to extract power from the atom in the year 1700. On the whole, the development of the technology of production presents a fairly smooth and continuous profile rather than one of jagged peaks and discontinuities.

The Predictability of Technology

There is a long history of technological prediction, some of it ludicrous and some not.[4] What is interesting is that the development of technical progress has always seemed *intrinsically* predictable. This does not mean that we can lay down future timetables of technical discovery, nor does it rule out the possibility of surprises. Yet I venture to state that many scientists would be willing to make *general* predictions as to the nature of technological capability 25 or even 50 years ahead. This, too, suggests that technology follows a developmental sequence rather than arriving in a more chancy fashion.

I am aware, needless to say, that these bits of evidence do not constitute anything like a "proof" of my hypothesis. At best they establish the grounds on which a *prima facie* case of plausibility may be rested. But I should like now to strengthen these grounds by suggesting two deeper-seated reasons why technology *should* display a "structured" history.

The first of these is that a major constraint always operates on the technological capacity of an age, the constraint of its accumulated stock of available knowledge. The application of this knowledge may lag behind its reach; the technology of the hand-mill, for example, was by no means at the frontier of medieval technical knowledge, but technical realization can hardly precede what men generally know (although experiment may incrementally advance both technology and knowledge concurrently). Particularly from the mid-nineteenth century to the present do we sense the loosening constraints on technology stemming from successively yielding barriers of scientific

4. Jewkes et al. (see n. 2) present a catalogue of chastening mistakes (p. 230f.). On the other hand, for a sober predictive effort, see Francis Bellow, "The 1960s: A Forecast of Technology," *Fortune* 59 (January 1959): 74–78; Daniel Bell, "The Study of the Future," *Public Interest* 1 (fall 1965): 119–130. Modern attempts at prediction project likely avenues of scientific advance or technological function rather than the feasibility of specific machines.

knowledge—loosening constraints that result in the successive arrival of the electrical, chemical, aeronautical, electronic, nuclear, and space stages of technology.[5]

The gradual expansion of knowledge is not, however, the only order-bestowing constraint on the development of technology. A second controlling factor is the material competence of the age, its level of technical expertise. To make a steam engine, for example, requires not only some knowledge of the elastic properties of steam but the ability to cast iron cylinders of considerable dimensions with tolerable accuracy. It is one thing to produce a single steam machine as an expensive toy, such as the machine depicted by Hero, and another to produce a machine that will produce power economically and effectively. The difficulties experienced by Watt and Boulton in achieving a fit of piston to cylinder illustrate the problems of creating a technology, in contrast with a single machine.

Yet until a metal-working technology was established—indeed, until an embryonic machine-tool industry had taken root—an industrial technology was impossible to create. Furthermore, the competence required to create such a technology does not reside alone in the ability or inability to make a particular machine (one thinks of Babbage's ill-fated calculator as an example of a machine born too soon), but in the ability of many industries to change their products or processes to "fit" a change in one key product or process.

This necessary requirement of technological congruence[6] gives us an additional cause of sequencing. For the ability of many industries to cooperate in producing the equipment needed for a "higher" stage of technology depends not alone on knowledge or sheer skill but on the division of labor and the specialization of industry. And this in turn hinges to a considerable degree on the sheer size of the stock of capital itself. Thus the slow and painful accumulation of capital,

5. To be sure, the inquiry now regresses one step and forces us to ask whether there are inherent stages for the expansion of knowledge, at least insofar as it applies to nature. This is a very uncertain question. But having already risked so much, I will hazard the suggestion that the roughly parallel sequential development of scientific understanding in those few cultures that have cultivated it (mainly classical Greece, China, the high Arabian culture, and the West since the Renaissance) makes such a hypothesis possible, provided that one looks to broad outlines and not to inner detail.

6. The phrase is Richard LaPiere's in *Social Change* (New York, 1965), p. 263f.

from which springs the gradual diversification of industrial function, becomes an independent regulator of the reach of technical capability.

In making this general case for a determinate pattern of technological evolution—at least insofar as that technology is concerned with production—I do not want to claim too much. I am well aware that reasoning about technical sequences is easily faulted as *post hoc ergo propter hoc.* Hence, let me leave this phase of my inquiry by suggesting no more than that the idea of a roughly ordered progression of productive technology seems logical enough to warrant further empirical investigation. To put it as concretely as possible, I do not think it is just by happenstance that the steam-mill follows, and does not precede, the hand-mill, nor is it mere fantasy in our own day when we speak of the coming of the automatic factory. In the future as in the past, the development of the technology of production seems bounded by the constraints of knowledge and capability and thus, in principle at least, open to prediction as a determinable force of the historical process.

II

The second proposition to be investigated is no less difficult than the first. It relates, we will recall, to the explicit statement that a given technology imposes certain social and political characteristics upon the society in which it is found. Is it true that, as Marx wrote in *The German Ideology,* "A certain mode of production, or industrial stage, is always combined with a certain mode of cooperation, or social stage,"[7] or that, as he put it in the sentence immediately preceding our hand-mill–steam-mill paradigm, "in acquiring new productive forces men change their mode of production, and in changing their mode of production they change their way of living—they change all their social relations"?

As before, we must set aside for the moment certain "cultural" aspects of the question. But if we restrict ourselves to the functional relationships directly connected with the process of production itself, I think we can indeed state that the technology of a society imposes a determinate pattern of social relations on that society.

7. Karl Marx and Friedrich Engels, *The German Ideology* (London, 1942), p. 18.

We can, as a matter of fact, distinguish at least two such modes of influence:

The Composition of the Labor Force
In order to function, a given technology must be attended by a labor force of a particular kind. Thus, the hand-mill (if we may take this as referring to late medieval technology in general) required a work force composed of skilled or semiskilled craftsmen, who were free to practice their occupations at home or in a small atelier, at times and seasons that varied considerably. By way of contrast, the steam-mill—that is, the technology of the nineteenth century—required a work force composed of semiskilled or unskilled operatives who could work only at the factory site and only at the strict time schedule enforced by turning the machinery on or off. Again, the technology of the electronic age has steadily required a higher proportion of skilled attendants; and the coming technology of automation will still further change the needed mix of skills and the locale of work, and may as well drastically lessen the requirements of labor time itself.

The Hierarchical Organization of Work
Different technological apparatuses require not only different labor forces but different orders of supervision and coordination. The internal organization of the eighteenth-century handicraft unit, with its typical man-master relationship, presents a social configuration of a wholly different kind from that of the nineteenth-century factory with its men-manager confrontation, and this in turn differs from the internal social structure of the continuous-flow, semi-automated plant of the present, as the intricacy of the production process increases, a much more complex system of internal controls is required to maintain the system in working order.

Does this add up to the proposition that the steam-mill gives us society with the industrial capitalist? Certainly the class characteristics of a particular society are strongly implied in its functional organization. Yet it would seem wise to be very cautious before relating political effects exclusively to functional economic causes. The Soviet Union, for example, proclaims itself to be a socialist society although its technical base resembles that of old-fashioned capitalism. Had Marx written that the steam-mill gives you society with the industrial *manager*, he would have been closer to the truth.

What is less easy to decide is the degree to which the technological infrastructure is responsible for some of the sociological features of society. Is anomie, for instance, a disease of capitalism or of all industrial societies? Is the organization man a creature of monopoly capital or of all bureaucratic industry wherever found? These questions tempt us to look into the problem of the impact of technology on the existential quality of life, an area we have ruled out of bounds for this paper. Suffice it to say that superficial evidence seems to imply that the similar technologies of Russia and America are indeed giving rise to similar social phenomena of this sort.

As with the first portion of our inquiry, it seems advisable to end this section on a note of caution. There is a danger, in discussing the structure of the labor force or the nature of intrafirm organization, of assigning the sole causal efficacy to the visible presence of machinery and of overlooking the invisible influences of other factors at work. Gilfillan, for instance, writes, "engineers have committed such blunders as saying the typewriter brought women to work in offices, and with the typesetting machine made possible the great modern newspaper, forgetting that in Japan there are women office workers and great modern newspapers getting practically no help from typewriters and typesetting machines."[8] In addition, even where technology seems unquestionably to play the critical role, an independent "social" element unavoidably enters the scene in the *design* of technology, which must take into account such facts as the level of education of the work force or its relative price. In this way the machine will reflect, as much as mold, the social relationships of work.

These caveats urge us to practice what William James called a "soft determinism" with regard to the influence of the machine on social relations. Nevertheless, I would say that our cautions qualify rather than invalidate the thesis that the prevailing level of technology imposes itself powerfully on the structural organization of the productive side of society. A foreknowledge of the shape of the technical core of society 50 years hence may not allow us to describe the political attributes of that society, and may perhaps only hint at its sociological character, but assuredly it presents us with a profile of requirements, both in labor skills and in supervisory needs, that differ considerably from those of today. We cannot say whether the society

8. Gilfillan (see n. 3), p. 202.

of the computer will give us the latter-day capitalist or the commissar, but it seems beyond question that it will give us the technician and the bureaucrat.

III

Frequently, during our efforts thus far to demonstrate what is valid and useful in the concept of technological determinism, we have been forced to defer certain aspects of the problem until later. It is time now to turn up the rug and to examine what has been swept under it. Let us try to systematize our qualifications and objections to the basic Marxian paradigm:

Technological Progress Is Itself a Social Activity

A theory of technological determinism must contend with the fact that the very activity of invention and innovation is an attribute of some societies and not of others. The Kalahari bushmen or the tribesmen of New Guinea, for instance, have persisted in a neolithic technology to the present day; the Arabs reached a high degree of technical proficiency in the past and have since suffered a decline; the classical Chinese developed technical expertise in some fields while unaccountably neglecting it in the area of production. What factors serve to encourage or discourage this technical thrust is a problem about which we know extremely little at the present moment.[9]

The Course of Technological Advance Is Responsive to Social Direction

Whether technology advances in the area of war, the arts, agriculture, or industry depends in part on the rewards, inducements, and incentives offered by society. In this way the direction of technological advance is partially the result of social policy. For example, the system of interchangeable parts, first introduced into France and then independently into England, failed to take root in either country for lack of government interest or market stimulus. Its success in America is attributable mainly to government support and to its appeal in a

9. An interesting attempt to find a line of social causation is found in E. Hagen, *The Theory of Social Change* (Homewood, Ill., 1962).

society without guild traditions and with high labor costs.[10] The general *level* of technology may follow an independently determined sequential path, but its areas of application certainly reflect social influences.

Technological Change Must be Compatible with Existing Social Conditions

An advance in technology not only must be congruent with the surrounding technology but must also be compatible with the existing economic and other institutions of society. For example, labor-saving machinery will not find ready acceptance in a society where labor is abundant and cheap as a factor of production. Nor would a mass-production technique recommend itself to a society that did not have a mass market. Indeed, the presence of slave labor seems generally to inhibit the use of machinery and the presence of expensive labor to accelerate it.[11]

These reflections on the social forces bearing on technical progress tempt us to throw aside the whole notion of technological determinism as false or misleading.[12] Yet, to relegate technology from an undeserved position of *primum mobile* in history to that of a mediating factor, both acted upon by and acting on the body of society, is not to write off its influence but only to specify its mode of operation with greater precision. Similarly, to admit we understand very little of the cultural factors that give rise to technology does not depreciate its role but focuses our attention on that period of history when technology is clearly a major historical force, namely Western society since 1700.

IV

What is the mediating role played by technology within modern Western society? When we ask this much more modest question, the interaction of society and technology begins to clarify itself for us.

10. See K. R. Gilbert, "Machine-Tools," in *A History of Technology,* ed. C. Singer et al. (Oxford, 1958), IV, chapter xiv.

11. See LaPiere (n. 6), p. 284; also H. J. Habbakuk, *British and American Technology in the 19th Century* (Cambridge, 1962), passim.

12. As, for example, in A. Hansen, "The Technological Determination of History," *Quarterly Journal of Economics* (1921): 76–83.

The Rise of Capitalism Provided a Major Stimulus for the Development of a Technology of Production

Not until the emergence of a market system organized around the principle of private property did there also emerge an institution capable of systematically guiding the inventive and innovative abilities of society to the problem of facilitating production. Hence the environment of the eighteenth and nineteenth centuries provided both a novel and an extremely effective encouragement for the development of an *industrial* technology. In addition, the slowly opening political and social framework of late mercantilist society gave rise to social aspirations for which the new technology offered the best chance of realization. It was not only the steam-mill that gave us the industrial capitalist but the rising inventor-manufacturer who gave us the steam-mill.

The Expansion of Technology within the Market System Took on a New "Automatic" Aspect

Under the burgeoning market system not alone the initiation of technical improvement but its subsequent adoption and repercussion through the economy was largely governed by market considerations. As a result, both the rise and the proliferation of technology assumed the attributes of an impersonal diffuse "force" bearing on social and economic life. This was all the more pronounced because the political control needed to buffer its disruptive consequences was seriously inhibited by the prevailing laissez-faire ideology.

The Rise of Science Gave a New Impetus to Technology

The period of early capitalism roughly coincided with and provided a congenial setting for the development of an independent source of technological encouragement—the rise of the self-conscious activity of science. The steady expansion of scientific research, dedicated to the exploration of nature's secrets and to their harnessing for social use, provided an increasingly important stimulus for technological advance from the middle of the nineteenth century. Indeed, as the twentieth century has progressed, science has become a major historical force in its own right and is now the indispensable precondition for an effective technology.

It is for these reasons that technology takes on a special significance in the context of capitalism—or, for that matter, of a socialism based on maximizing production or minimizing costs. For in these societies, both the continuous appearance of technical advance and its diffusion throughout the society assume the attributes of autonomous process, "mysteriously" generated by society and thrust upon its members in a manner as indifferent as it is imperious. This is why, I think, the problem of technological determinism—of how machines make history—comes to us with such insistence despite the ease with which we can disprove its more extreme contentions.

Technological determinism is thus peculiarly a problem of a certain historical epoch—specifically that of high capitalism and low socialism—*in which the forces of technical change have been unleashed, but when the agencies for the control or guidance of technology are still rudimentary.*

The point has relevance for the future. The surrender of society to the free play of market forces is now on the wane, but its subservience to the impetus of the scientific ethos is on the rise. The prospect before us is assuredly that of an undiminished and very likely accelerated pace of technical change. From what we can foretell about the direction of this technological advance and the structural alterations it implies, the pressures in the future will be toward a society marked by a much greater degree of organization and deliberate control. What other political, social, and existential changes the age of the computer will also bring we do not know. What seems certain, however, is that the problem of technological determinism—that is, of the impact of machines on history—will remain germane until there is forged a degree of public control over technology far greater than anything that now exists.

Technological Determinism Revisited

Robert Heilbroner

"Do Machines Make History?" appeared in 1967 in *Technology and Culture*. Probably written a year earlier, it has now attained senescence by the standards of life expectancy of journal articles. I must confess that I had assumed it had long ago descended into the Great Limbo that awaits us all. All the greater pleasure, then, to see the article referred to as classic, although these days one is not certain if that refers to the standards of Cicero or Coca-Cola; more important, all the better reason to reassess its findings, now that they and I are both a quarter of a century older.

The article is an attempt to examine the idea of technological determinism as a powerful force of history—especially the history of large-scale socioeconomic transformations, of which the most important are the transition from feudalism to capitalism and the evolution of capitalism through its various stages. A great deal of attention is therefore devoted to the means by which "force" is generated in the flow of events through changes in the material basis of social life, and to the kinds of changes that this force effects. Hence, much of the original article is preoccupied with the problems of describing how technology evolves and with the linkages that connect it to social change.

These general themes still strike me as constituting the analytical core of the idea of technological determinism, and I shall accordingly devote a small part of this revisitation to a few emendations of and additions to my earlier conclusions. These changes are not of great significance—not because a rereading of my piece convinces me that there is nothing further of interest to be said on the topic, but simply because I cannot think of it. Perhaps for that reason, the aspect that today attracts me to the subject is one that 25 years ago had not yet caught my attention. It focuses on the question of what we take "history" to be, and why we are drawn to a technological mode of interpreting its palimpsest.[1]

That is only metaphor. The substantive problem is why we might be receptive to such a way of construing history. Here I must risk a generalization. It is that the attribute of "modern" historiography that most sharply distinguishes it from "premodern" (I use the terms to refer to styles, not to periods) is its treatment of the background panorama against which the foreground inquiry takes place. In premodern history, that panorama is dominated by inscrutable forces. In the foreground pharaohs and emperors come and go, city-states rise and fall, good kings follow bad and vice versa, but all the while, behind these adventures and misadventures we espy forces that obey a different causality—the whims of the gods, the dramas of cosmology and salvationism, or simply the intervention of chance, luck, and the like. "The only thing," writes Collingwood, "that a shrewd and critical Greek like Herodotus would say about the divine power that ordains the course of history is that . . . it rejoices in upsetting and disturbing things."[2]

In sharp contrast, the basic premise of modern historiography is that background forces arise from the same kinds of processes, and can be approached and apprehended at the same levels of understanding and explanation, as the objects of immediate scrutiny. This unification of foreground and background is perhaps most strikingly evident in the rise of an "economic" interpretation of history. From its initial eighteenth-century formulations (I think of Adam Ferguson's *Essay on Civil Society*) to Marxian materialism, Braudelian stratification, or neoclassical choice theory, the hallmark of modern historical work is an effort to establish filiations between the subject that has been singled out for treatment and the background against which the subject is displayed. Wars and political events, the staple

1. My dictionary (*Webster's New Twentieth Century,* 1971) tells me that a palimpsest is a "parchment or tablet that has been written upon or inscribed two or three times, the previous text or texts having been imperfectly erased and remaining, therefore, still visible." That is not a bad description of the writing of history. Technological determinism then becomes a prescription for grinding our lenses, enabling us to make out characters on the parchment that have not previously been seen or understood and thereby to write new texts or to read old ones in new ways.

2. R. G. Collingwood, *The Idea of History* (Oxford University Press, 1956), p. 22.

subjects of premodern history, are now of interest largely insofar as they embody and concretize background forces such as class struggle and rational maximizing. Indeed, it is characteristic of modern history that the "subject" becomes these very background forces themselves, and that individual figures or events are studied not so much for their intrinsic interest as to illustrate or instantiate the larger processes that interest the historian.

It is here, of course, that technological determinism enters the picture. Indeed, because modern historiography takes for granted that technology, like all background elements, must perforce penetrate the narratives of history-writing, the question changes from "Do machines make history?" to "How do machines make history?"—a change that opens the way for definitions, boundaries, and argument rather than for polemics or declarations of faith.

Technology and Material Life

Let us then look at some of the ways in which machines make history on the grand scale that interests us here. One such way comes immediately to mind: Machines make history by changing the material conditions of human existence. It is largely machines (here I use the term to denote both individual mechanisms and a general level of technological development) that define what it means to live in a certain epoch—at least, as an economic historian might define life. Elsewhere in this volume, Rosalind Williams points out that such an economics-oriented viewpoint not only begs serious questions but also establishes powerful agendas. I shall deal with that problem before I am done, but it is useful to begin by considering the intimate and pervasive engagement of machinery with everyday life.

A paradoxical aspect of this interconnection is that the engagement is, at the same time, the most immediately apparent example of how machines make history and the least satisfactory example of what we might mean by technological determinism. I shall spend only a few words on the first part of this assertion. If we wish to study a society unfamiliar to us, the best place to start is by grasping its material life. To understand the historical significance of Eileen Power's peasant Bodo, of Mantoux's Arkwright, or of Marx's Moneybags we must first become acquainted with the material circumstances of their lives. In the same way, an understanding of the events of contemporary history takes for granted a knowledge of the technological setting that shapes modern-day existence as the mountains and the sea shaped life in the premodern Mediterranean.

Yet such study does not answer the question of how machines make history. A recognition that the technological structure is inextricably entwined in the activities of any society does not shed light on the connection between changes in that structure and changes in the socioeconomic order.[3] Of course we would expect that the transition from feudalism to capitalism was profoundly connected with the rise of a new level of technological capability. But precisely what does "profoundly" mean? That is the question to which technological determinism promises to yield an elucidation, which is perhaps the closest thing to an answer that the question permits. Lacking such an elucidation, we cannot give any kind of *analytic* account of the manner in which changes in machines alter daily life. Hence, we may be tempted to depict the meaning of technological determinism as the ascription to machines of "powers" they do not have. This leads to impressive-sounding but ultimately unsupportable statements, such as Veblen's assertion that "the machine throws out anthropomorphic habits of thought" and Marx's expectation that the introduction of the railway into India would "dissolve" the caste system.[4]

The challenge, then, is to demonstrate that technology exerts its effects in generalizable ways. If technological determinism is to become a useful overlay for history's palimpsest, it must reveal a connection between "machinery" and "history" that displays lawlike properties—a force field, if we will, emanating from the technological background to impose order on human behavior in a manner anal-

3. The recognition does, however, raise the important question of whether technology is itself a background or a foreground element. The answer depends on what we are seeking to investigate. If our interest lies in the sources of technological change itself, as in Joel Mokyr's book *The Lever of Riches* (Oxford University Press, 1990), technology becomes the foreground element on whose development impinge forces emanating from the socioeconomic background. If the dynamics of social formations themselves lie at the focal point of inquiry, the placements are reversed, and we seek causal factors in the technological background. It is the latter perspective that is normally associated with technological determinism.

4. Thorstein Veblen, *The Theory of Business Enterprise* (Scribners, 1932), p. 310. Marx from Michael Adas, *Machines as the Measure of Men: Science, Technology, and Ideologies of Western Dominance* (Cornell University Press, 1989), p. 240.

ogous to that by which a magnet orders the behavior of particles sprinkled on a sheet of paper held above it or that by which gravitation orders the paths of celestial objects.

Is there such a force field? It is not difficult to find candidates for its order-bestowing task. Whitehead credited "routine," without which "civilization vanishes." Many have located the source of behavioral regularity in "human nature," variously described: Hume said that human motivation was a kind of repeating decimal in history.[5] For our purposes, what is lacking in these principles is an ability to translate the stimuli emitted by a changing technology into behavioral responses of a predictable kind with regard to transformations of a socioeconomic order. Routine will not serve that purpose; it tells us only that technological change will be resisted, not the form that resistance will take. Human nature fails for the same reason, even when it is reduced to its constitutive drives—aggression, self-preservation, or whatever—because we have no way of generalizing about the effects of technological change on these drives or about the effects of changes in the drives on the social framework.

What is needed, in other words, is a mechanism of a near-alchemical kind. A huge variety of stimuli, arising from alterations in the material background, must be translated into a few well-defined behavioral vectors. There must be a systematic reduction of complexity of cause into simplicity of effect, enabling us to explain how the development of new machineries of production can alter the social relationships constitutive of feudalism into those of capitalism, or those of one kind of capitalism into those of another kind. Such a mechanism seems impossible to imagine. What is perhaps even more imagination-defying is that it exists.

Economics as Force Field

The mechanism is, of course, economics, in the sense of a force field in which a principle of "maximizing" imposes order on behavior in a fashion comparable to the magnet and the gravitational pull of the sun. I shall consign to a note the meaning of the vexed term I have put into quotes. For our purposes, it will be sufficient to describe it in Adam Smith's straightforward formulation: "bettering our condi-

5. Alfred North Whitehead, *Adventures of Ideas* (Macmillan, 1933), p. 113; Hume, *An Enquiry Concerning Human Understanding*, section V, part 1.

tion [by] an augmentation of fortune."[6] This rough-and-ready description, for which "acquisitive mindset" is perhaps a more precise equivalent, is enough to reduce the varied stimuli of changes in the material environment to well-specified behavioral results.

At the risk of belaboring the obvious, economics accomplishes this remarkable feat by ignoring all effects of the changed environment except those that affect our maximizing possibilities. In this way, changes in technology, like changes in the weather or in our social situations, are depicted as loosenings or tightenings of constraints on our behavior, and these altered constraints are then perceived as changing our actions in sufficiently regularized ways to enable us to speak of "laws" at work in the marketplace or in the enterprise.

It is not necessary to discuss whether or not this regularized behavioral response can be considered a part of human nature.[7] It is enough that the acquisition of fortune becomes a widely noted "rule" of behavior in societies that leave behind the coordinating mechanisms of tradition and command for those of the market. Thus far in history these societies are, in fact, all members of the general social formation we call capitalism. It follows that economic determinism, and its technological correlate, have relevance only in the capitalist social order, in which, as Marx emphasized, the multifarious world of use values is transmuted into a one-dimensional world of exchange values.

In this social order, changes in the technological background are registered in changes in the price system, indicating the directions in which economic activity can most advantageously move and the forms it can most profitably assume. Thus the force field of maximizing allows us to elucidate how machines make history by showing the

6. Adam Smith, *The Wealth of Nations* (Modern Library, 1936), pp. 324–325. Maximizing described in terms of utility does not yield an action-directive capable of serving as an operational force field, insofar as utility maximization is tautologous with respect to behavior. Only a Smithian maximization of "fortune" provides the required specificity of response. Fortunately, this appears to be a reasonable description of economic life, most departures from the rule being self-correcting.

7. For support of this proposition see Jack Hirschleifer, "The Expanding Domain of Economics," *American Economic Review* (December 1985): 53, or Gary Becker, *The Economic Approach to Human Behavior* (University of Chicago Press, 1976). For a critique see chapters 1 and 7 of my own *Behind the Veil of Economics* (Norton, 1988).

mediating mechanism by which changes in technology are brought to bear on the organization of the social order. This leads in turn to the possibility of applying an analytic understanding to such large-scale social changes as the composition of the labor force and the hierarchical organization of work, not to mention the dynamics characteristic of economic activity as a whole.

Determinism as Heuristic

This immediately raises many questions and some specters. To begin, the triadic connection of technological determinism, economic determinism and capitalism does not mean that technology has no effects on noncapitalist society. The stirrup exerted a "catalytic" impact on the socioeconomic organization of Carolingian society, as did the spread of iron weapons from 1200 to 800 B.C. and the advent of printing in the fifteenth century. The difference is that precapitalist technological impingements do not affect their societies with the "logic" that comes only with capitalism's translation of use values into exchange values. The impact of precapitalist technical change therefore appears more contingent, less open to systematic elucidation, than when an economic force field guides its applications and consequences. We can say many more things about the "path" of technical change in the United States in the nineteenth century than about its course in ancient China or the Roman empire.[8] Perhaps there are other logics that would enable us to describe the interaction of technical change and social consequence with the degree of predictability and precision that is the very identifying characteristic of economics, but we do not know them.

Next, I hasten to add that in elevating economic determinism to the rank of a tutelary deity within capitalism, I am not asserting that economics is thereby entitled to assume the rank of the queen of the social sciences. I believe that what we call economic behavior is best understood as the sublimated expression of much deeper-rooted elements of "political" and "social" behavior—dominance and obedience—which can, in turn, be traced to the human experience of

8. Compare Ross Thompson, *The Path to Mechanized Shoe Production in the United States* (University of North Carolina, 1989), and chapters 2 and 9 of Mokyr, *The Lever of Riches.*

protracted childhood dependency.[9] Nonetheless, what is important is that "economic" behavior—that is, behavior motivated by the pursuit of exchange value—is set analytically apart from and above behavior motivated by other considerations, because it manifests a degree of orderliness absent from "political" or "sociological" activities. It may be that politics and sociology contain their own laws—I think of Alphonse Karr's "Plus ça change, plus c'est la même chose" and Lord Acton's pronouncements about the corrupting tendencies of power. Unlike economic behavior, however, these generalizations do not give us the translation mechanism of economics, enabling us—to take the present instance—to speak systematically about how machines will bring pressures to bear on socioeconomic formations.

Finally, technological determinism does not imply that human behavior must be deprived of its core of consciousness and responsibility. It avoids this trap insofar as it offers a heuristic of investigation, not a logic of decision-making. As such a heuristic, it presents a premise from which we can initially approach the interpretation of socioeconomic events, not an infamous "last instance" by which we are forced ultimately to resolve it. The premise is that living behavior is not random or chaotic, but marked by undeniable, although imprecise regularities: "undeniable" because predictable behavior is the basis on which all social life is raised; "imprecise" not only because there are obvious variances in behavior from one social entity to the next, but also because there exists a margin of behavioral indeterminacy within any given individual. Technological determinism then goes on to posit that the acquisitive mindset is a regular and dependable motive for behavior, at least in market-coordinated societies. Combining this persistent general drive and the margins of freedom characteristic of its concrete expression, we arrive at the formulation of a "soft determinism" which, however paradoxical to the philosopher, should present no great difficulties to the historian.

Degrees of Determinism

It is time to place these rather abstract considerations into more concrete perspective. Alfred Chandler's recent survey of American, British, and German capitalism during the period 1919–1948 pro-

9. Robert Heilbroner, *The Nature and Logic of Capitalism* (Norton, 1985), pp. 46–52.

vides a useful case in point.[10] Chandler's immediate interest is directed to explaining how the structures of capitalism were shaped by managerial styles that he describes as Competitive (American), Personal (British), and Cooperative (German). These styles, in turn, are shown to reflect underlying sociopolitical elements—what we might be tempted to call the respective "national characters" of the three nations. Chandler's important contribution is to point out how these differing responses to a common problem of industrial dynamics affected the developmental paths (and today, the development prospects) of the three nations.

This is an enlightening means of exploring the relation of technological and economic determinism, and the degrees of hardness and softness of these order-bestowing principles, although I hasten to add that these are my characterizations and not Chandler's. Chandler shows how the introduction of high-speed, continuous-run, capital-intensive machines led to a number of quite different, even mutually contradictory, institutional changes. Here is where technological determinism supplies many elucidations. The first is that the physical characteristics of the machinery of mass production can be discerned as the material cause of the phenomenon that Chandler is examining. These characteristics are not, however, the efficient cause. That lies in the translation of the engineering consequences of mass production into the economic stimuli of large changes in cost per unit of production—a translation that makes visible the force field of maximization to which activity is exposed in the market sphere of capitalism.

This is the "deterministic" aspect of the elucidation, and if that were all there were to it, we would expect that managements in all industrial capitalisms would respond in similar fashion. Now the equally important soft considerations enter, for the dramatic economies of scale and scope can be interpreted as opening new possibilities for—or new dangers from—corporate expansion. As Chandler shows, this makes way for a range of institutional responses, each a maximizing answer to the perceived economic situation. In this way, softness combines with hardness to throw analytic light on the different ways in which machines can influence the socioeconomic

10. Alfred D. Chandler, Jr., *Scale and Scope: The Dynamics of Industrial Capitalism* (Harvard University Press, 1990).

framework, even though all are obeying a common economic imperative.

Needless to say, this view of technological determinism is far removed from any kind of mechanical linkage. For instance, even where the specific characteristics of technology give rise to well-defined matrices of inter-industry connections, these linkages may not guide the course of economic growth if certain "soft" political and social preconditions are absent. Marx seems to have overlooked this possibility in speaking about the consequences of railroadization for India:

> . . . when you have once introduced machinery into the locomotion of a country, which possesses iron and coal, you are unable to withhold it from its fabrication. You cannot maintain a net of railways over an immense country without introducing all those industrial processes necessary to meet the immediate and current wants of railway locomotion and out of which there must grow the application of machinery to those branches of industry not immediately connected with railways. The railway system will become, in India, truly the forerunner of modern industry.[11]

Here Marx is talking about the railroad as an apparatus whose backward and forward linkages would eventually force a leap from one level of technological capability to another. Yet historical experience, not least that of India, has shown us that the logic of industrial linkage loses its peremptory force in determining the developmental path of a nation entangled in a preexisting global division of industry. The failure of centrally planned economies to duplicate the performance of capitalist economies is perhaps even more illustrative of such soft considerations. The history of Soviet industrialization has shown that technology is the servant and not the master of its associated system of sociopolitical directives, and that the use of more or less identical kinds of lathes, presses, or assembly-line configurations in centralized planning systems and in capitalist market systems will not result in anything like identical growth paths or performance levels.[12]

Indeed, if there is any substantial recantation that 25 years of observation has enjoined on me, it is to withdraw my earlier inclusion of "low socialism" along with high capitalism as a setting in which it would be possible to explain, even to foresee, the broad socioeco-

11. From Adas, *Machines as the Measure of Men,* p. 240.

12. See Nikolai Shmelev and Vladimir Popov, *The Turning Point: Revitalizing the Soviet Economy* (Doubleday, 1989).

nomic consequences flowing from changes in the machinery of production. I had assumed that low socialism meant a society that still embraced—and was still embraced by—capitalist-like motivations. In the absence of that premise, the application of technological—and, of course, of economic—determinism to the developmental tendencies of any form of socialism seems extremely uncertain.

The Bed of Procrustes

After so much has been taken away, what remains of technological determinism as a perspective from which we can consider how machines make history, always bearing in mind the socioeconomic meaning we attach to the last word in the question?

- Technological determinism gives us a framework of explication that ties together the background forces of our civilization, in which technology looms as an immense presence, with the foreground problem of the continuously evolving social order in which we live. The tie between the two is far from definitive, complete, or unambiguous, but it is the only such connection that we can make. As such, it has deep appeals to the modern temper, which recoils from introducing unknowability into the course of events. The deterministic view does not foreclose on a margin of indeterminacy (in some cases a very large margin), but its active search for regularities in, and lawlike aspects to, historical change remains the most powerful unifying capability we have. Whatever compression we may thereby suffer in our conceptions of ourselves as actors in history seems to me far less than would be felt if the very idea of a historic orderliness were shown to be utterly without basis. History as contingency is a prospect that is more than the human spirit can bear.

- Even in the most dramatic instances of technological determinism, as when we can trace the socioeconomic effects of the factory, the technique of mass production, or the modern-day computer, we can never eliminate the soft causal elements that are always present with, and within, those of the economic force field itself. Among these soft elements we must place many volitional elements, including most of what we call political decisions, social attitudes, cultural fads and fashions, and those aspects of maximizing itself in which the agent's final determination hinges on time horizons, risk aversion, and similar judgments about which no behavioral generalizations can be made. Hence the clarifying power of determinism, even at its greatest, must always allow for some degree of uncertainty. This is perhaps only tantamount to saying that our conceptions of "history" cannot embrace either a fully determined or a wholly undetermined narrative of events—a state of affairs that no doubt reveals more about our psychological limitations than about the actualities of historical sequence, whatever they may be.

• Rosalind Williams protests that a deterministic heuristic, however soft, involves us in a cramping view of history's palimpsest, throwing aside much of its endlessly rich interpretational receptivity in favor of the thin stuff of behavioral "laws" and socioeconomic "formations."[13] I am very sympathetic to her denunciation of economism as a Procrustean approach to historical understanding, and no friend of the pretensions of economics to be a "universal science."[14] Disapproval is one thing, however; disavowal another. We live in a social order in which an economic calculus takes precedence over, and enters into the definitions of, many aspects of life. In the land of Procrustes, the standards of the innkeeper are those that apply. As long as economics constitutes the most powerful and pervasive motivational force, and the only one to which behavioral regularities can be ascribed, a perspective of soft determinism seems to me the one most likely to enable us to grasp the processes of history in which we are entangled.

Acknowledgement

I would like to thank Ross Thompson for his extremely valuable criticism.

13. These are my terms of derogation, but I think they catch her drift.

14. See my "Analysis and Vision in the History of Modern Economic Thought," *Journal of Economic Literature* (September 1990): 1111–1113, and "Economics as Ideology," in *Economics as Discourse*, ed. W. J. Samuels (Kluwer-Nijhoff, 1990).

Three Faces of Technological Determinism

Bruce Bimber

In this essay Bruce Bimber takes a different tack than Heilbroner. Specifically, he distinguishes three interpretations of technological determinism—the Normative, Nomological, and Unintended Consequences accounts—and develops a set of criteria with which to test the rigor and the legitimacy of these interpretations. Among other things, Bimber's strict approach "does not admit such distinctions as 'hard' and 'soft' determinism." He argues that only nomological accounts are "truly technologically deterministic." He tests his model by applying it to the work of Karl Marx and shows that Marx was not a technological determinist. "For Marx," he concludes, "technology was no more than one kind of important and efficient fuel for history's human engine."

The idea that technological development determines social change has a remarkably tenacious grip on the popular and the academic imagination. In spite of the best efforts of historians and others to show that the relationships between technology and society are reciprocal rather than unidirectional, claims for the autonomous influence of technology on societies persist. Technological determinism seems to lurk in the shadows of many explanations of the role of technology in human history.

Probably the most important reason for the tenacity of technological determinism in the face of attempts to disprove or dismiss it lies in the fact that the concept has been so flexible. Even a cursory examination of the literature on technology and social change reveals that "technological determinism" is used to describe more than one phenomenon. Technological determinism exists in enough different incarnations that the label can easily be attached to a range of views.

A lack of precision about the meaning of "technological determinism" fuels debates of all kinds about whether this concept accurately describes the unfolding of history. For example, whether Karl Marx was a technological determinist is a matter of perpetual debate among historians, economists, and philosophers. On one side in this debate are Robert Heilbroner, Langdon Winner, William Shaw, Alvin Hansen, and even Lewis Mumford, each of whom has placed the mantle of technological determinism on Marx in one way or another.[1] On the other side, Richard Miller, Nathan Rosenberg, Donald MacKenzie, and Reinhard Rürup have argued that Marx's historical materialism does not amount to technological determinism.[2]

1. Robert Heilbroner, "Do Machines Make History?" *Technology and Culture* 8 (winter 1967): 333–345 (reprinted in this volume); Langdon Winner, *Autonomous Technology* (MIT Press, 1977); William H. Shaw, "'The Handmill Gives You the Feudal Lord': Marx's Technological Determinism," *History and Theory* 18 (1979): 155–166; Alvin H. Hansen, "The Technological Interpretation of History," *Quarterly Journal of Economics* 36 (November 1921): 72–83; Lewis Mumford, *The Myth of the Machine: Technics and Human Development* (Harcourt Brace Jovanovich, 1966).

2. Richard Miller, *Analyzing Marx* (Princeton University Press, 1984); Nathan Rosenberg, "Marx as a Student of Technology," *Monthly Review* 28 (July–August 1976): 56–77; Donald MacKenzie, "Marx and the Machine," *Technology and Culture* 25 (June 1984): 473–502; Reinhard Rürup, "Historians and Modern Technology," *Technology and Culture* 15 (April 1974): 161–193; David Dickson, *The Politics of Alternative Technology* (Universe Books, 1975).

Lack of clarity about what is meant by "technological determinism" may lie behind much of this disagreement. Participants in the debate seem to be arguing more over the question of what is technological determinism than over what is Marxism. The same problem plagues many discussions of technology and historical change where technological determinism is invoked. Until we are able to agree about what precisely we mean by this concept, we are unlikely to resolve the question of whether technological determinism is a useful lens through which to interpret history.

I will return to the question of Marx, since, as Rosalind Williams points out in this volume, his spirit hovers over any discussion of this subject. My primary goal, however, is to identify the various concepts that go by the name "technological determinism." My chief claim is that the term is used in a muddy and imprecise way. Specifically, I argue that at least three distinct approaches to explaining historical change, reflecting quite different assumptions and offering quite different causal explanations, receive the label "technological determinism." I call these Normative accounts, Nomological accounts, and Unintended Consequences accounts. They range from positive descriptions of an inevitable technological order based on laws of nature (Nomological) to claims that technology is an important influence on history only where societies attach cultural and political meaning to it (Normative).

Beyond offering a typology of technological determinisms, I suggest a set of criteria for identifying which face of technological determinism is most meaningful and useful as an intellectual construct. In the second half of my discussion, I use the typology and the criteria to consider Marx's claims as a way of showing how a more precise view of technological determinism can sharpen and focus debate about its meaning and application in the study of history.

Technological Determinism

The most common interpretation of technological determinism is probably the least specific and precise, because it relies on human attitudes for explaining technology's historical influence. In *Toward a Rational Society,* Jürgen Habermas frames this interpretation with a study of technical enterprise.[3] Habermas is concerned with how soci-

3. Jürgen Habermas, *Toward a Rational Society* (Beacon, 1970), pp. 58–60.

eties can employ ethical conceptions to exert conscious, willful control over the norms of practice involved in technological development. He portrays this development as an essentially human enterprise in which the people who create and employ technology are driven by goals and judgments about public and private goods. Their actions follow certain culturally accepted norms and are sanctioned by politically legitimated forms of power.

Habermas's critique of this enterprise rests on the observation that industrial societies have developed an overreliance on norms of efficiency and productivity in directing the conduct of these processes. By adopting reductionistic values as guides to decision-making about technology, they exclude other ethical criteria, producing a self-correcting process that operates autonomous of larger political and ethical contexts. This social subsystem, made up of technologists pursuing the rationalization of life through the creation of technology, is autonomous when it frees itself from ethical, normative judgments directed at it from society at large. Complete acquiescence occurs when society has adopted as its own the technologists' standards of judgment.

Habermas's fear points us toward the essence of a first interpretation of technological determinism: norms of practice. Habermas suggests that technology can be considered autonomous and deterministic when the norms by which it is advanced are removed from political and ethical discourse and when goals of efficiency or productivity become surrogates for value-based debate over methods, alternatives, means, and ends. This is technological determinism's most familiar face, and I call this a Normative account.

In one way or another, the normative approach informs many incarnations of technological determinism. For many, the baldest statement of technological determinism is Jacques Ellul's book *The Technological Society*.[4] To Ellul, technique is not merely technology; it is the domination of social, political, and economic life by the adopted goals of logic and efficiency. As surrogates for value-laden norms and judgments, efficiency and technique lead to the technological society. A particular form of social practice, a set of overriding norms, produces this result. Ellul offers a Normative account of technological determinism.

4. Jacques Ellul, *The Technological Society* (Vintage, 1964).

This theme also runs through the work of several other prominent observers of technology and society. When Lewis Mumford warns us of the dangers of "megatechnics," when Herbert Marcuse describes the one-dimensional life of technological rationality, or when Langdon Winner describes "technology-out-of-control," each speaks to Habermas's concern.[5] These observers offer prescriptions against the adoption of a hegemonic cultural mindset that limits discourse and judgment to matters of efficiency. While Mumford, Marcuse, and Winner are themselves far from being technological determinists, they are often taken to be delivering the message that the advance of technology is autonomous and its influence deterministic.

The second interpretation of technological determinism is very different. It might be thought of as the analytic philosopher's version, because it adheres most closely to concepts in that discipline. The philosopher Peter Van Inwagen defines determinism as the belief that "given the past, and the laws of nature, there is only one possible future."[6] This general definition is the basis for what I refer to as the Nomological account of technology, which rests on laws of nature rather than on social norms. G. A. Cohen provides a useful discussion. Cohen takes technological determinism to mean that "machinery and allied subhuman powers somehow function as the independent agencies of history."[7] The claim here is that technology itself exercises causal influence on social practice. On Van Inwagen's definition, technological determinism can be seen as the view that, in light of the past (and current) state of technological development and the laws of nature, there is only one possible future course of social change. This might mean that various technological processes, once begun, require forms of organization or commitments of political resources, regardless of their social desirability or of previous social practices. It could mean that an enterprise (for example, the railroad) necessitates subsequent technologies (such as the telegraph,

5. Lewis Mumford, *The Pentagon of Power* (Harcourt Brace Jovanovich, 1964); Herbert Marcuse, *One-Dimensional Man* (Beacon, 1964); Langdon Winner, *Autonomous Technology* (MIT Press, 1977).

6. Peter Van Inwagen, *An Essay on Free Will* (Oxford University Press, 1983), p. 65.

7. G. A. Cohen, *Karl Marx's Theory of History: A Defense* (Princeton University Press, 1978), p. 147.

or large-scale, hierarchically organized steel-production facilities) and requires a pool of labor, the availability of capital, an insurance and banking industry, and so on, so that a fixed and predictable course of economic, social, and cultural change follows inevitably from the adoption of the railroad.

Richard Miller states this idea in a similar way. Miller understands technological determinism to mean that social structures evolve by adapting to technological change. That is, given a specific state of technology, "the subsequent development of society would be the same no matter what people thought or desired."[8]

Implicit in this account are two claims: that technological developments occur according to some naturally given logic, which is not culturally or socially determined, and that these developments force social adaptation and changes. Note how sharply this differs from Ellul's and Habermas's view. In Normative accounts, it is precisely that which people have come to think and desire that produces the technological society. In Nomological accounts, the technology-driven society emerges regardless of human desires and values.

The best-known Nomological interpretation of technology is probably Robert Heilbroner's classic article "Do Machines Make History?" Heilbroner describes a fixed sequence of technological developments prescribing the evolutionary path over which society must travel. He argues that "the steam-mill follows the hand-mill not by chance but because it is the next stage in a technical conquest of nature that follows one and only one grand avenue of advance." In the sense that this path or sequence is naturally given and independently drives social development, it is the primary causal agent in social change. History is predetermined by scientific laws that are sequentially discovered by people and which, in their inexorable application, produce technology. Only within the limits imposed by this logic may people exercise collective or individual agency and will. Heilbroner's grand avenue is therefore independent of culture and social practice. It is everywhere the same.[9]

8. Miller, *Analyzing Marx,* p. 183.

9. In his essay "Technological Determinism Revisited" (this volume), Heilbroner proposes a "soft" version of technological determinism that functions as a heuristic for interpreting capitalist economic history. That view departs from his original account in significant ways. At present, I refer only to Heilbroner's 1967 theory of history as a Nomological account.

Nomological accounts are, by definition, culture-independent, while Normative accounts are culture-specific. Although they rely on such different engines for explaining historical changes, they are easily conflated. John Staudenmaier acknowledges the presence of both kinds of interpretations, but he casts them as complementary premises. Staudenmaier claims that technological determinism is based on a disjunction between efficiency and other norms and on the advancement of technological progress in a fixed and necessary sequence.[10] He is correct to identify these two views; however, reconciling them into a single "determinist position" is misleading, because it mixes different causal explanations. Heilbroner's technological determinism is not compatible with Habermas's. Nomological accounts make a strong ontological claim, whereas Normative accounts make a cultural and attitudinal claim. These are distinct conceptions.

A third approach to explaining technology's role in social change, sometimes treated as technological determinism, focuses on the unanticipated effects of technological developments. Langdon Winner has described this view as an alternative to law-based interpretations of technology[11]; I label it the Unintended Consequences account, because it focuses on unintended consequences of technical enterprise. This account derives from observations of the uncertainty and uncontrollability of the outcomes of actions. For example, the early proponents of the automobile who argued that it would be environmentally cleaner than the old mode of transportation because it would rid the streets of horse manure did not foresee the environmental destruction that exhaust from the internal combustion engine would bring. This dramatic unintended consequence of the new technology emerged unsought and uncontrolled.[12]

The crux of this view is that even willful, ethical social actors are unable to anticipate the effects of technological development. For this reason, technology is at least partially autonomous. Technological developments have a role in determining social outcomes that is beyond human control. On the Nomological and Normative accounts, people might also be unable to anticipate the effects of technology;

10. John Staudenmaier, *Technology's Storytellers* (MIT Press, 1985), p. 136ff.

11. Winner, *Autonomous Technology*, p. 88ff.

12. James J. Flink, *The Car Culture* (MIT Press, 1975).

however, that fact is not necessary for either of those views. Those accounts posit more than simply unanticipated effects. Unintended Consequences accounts of technological determinism differ from Nomological accounts in that no particular set of laws—no discoverable underlying logic—necessarily drives the production of technology. And, unlike Normative accounts, Unintended Consequences accounts posit no specific cultural or social practice that produces the effects of technology. The focus is on the inability to know completely the consequences of technological choices rather than on the process by which technology is advanced.

The Three Accounts Compared

Normative, Nomological, and Unintended Consequences accounts provide three distinct interpretations of technological determinism. They attribute historical agency to quite different forces, based on models of social change that are at odds with one another. The differences among these views demand that we not speak of technological determinism as if it were a single concept or agreed-upon theory of history, for quite simply it is not.

Cohen has suggested two criteria for defining technological determinism that are useful for comparing these accounts. His gauge for the theory is attractive chiefly on grounds of semantic clarity and integrity. Cohen argues that technological determinism should nominally be both technological and deterministic. The first component of this view is that technological determinism should hold that history is determined by laws or by physical and biological conditions rather than by human will; this makes it deterministic. A generally accepted definition of determinism, apart from the study of technology, is the doctrine that future phenomena are causally determined by preceding events or natural laws. Cohen's first criterion merely holds technological determinism to this definition. By this rule, theories of history in which human agency is the primary cause of change are only obscured by the label " determinism." An important implication of this definition is that "determinism" should refer to covering laws of human action that apply at all times and all places in history. It cannot make sense to speak of determinism, understood in this way, as applying only in certain periods or as applying in various degrees.

The second component of Cohen's standard is that technological determinism should be truly technological in meaning. That is, technology should play a necessary part in the way that preceding events or states of the world determine the future. Most plausibly, this can be taken to mean that the laws of nature determining human history do so through technology. Technology is the medium through which physical laws, some of which we can learn through science, shape the course of human events. Only by meeting this standard can technological determinism differentiate itself from economic, theological, psychological, and other determinisms.[13] Cohen's standard insists that "technological determinism" not be used to describe accounts that are only deterministic, or only technological, or obviously neither. Perhaps the most basic requirement we can impose on a theory of technological determinism is that it be precise and distinct.

Requiring technological determinism to meet these criteria imposes a tough test indeed, because it rules out use of the term as a heuristic or as a catch-all label for conditions that may obtain temporarily in history. This standard does not admit of "hard" and "soft" versions of determinism, for a so-called soft determinism cannot be determinism at all.[14] In practice, "technological determinism" begins to lose its meaning when it is used as a malleable interpretive tool. There is much to be gained by forgoing the term as a shorthand description for the history of what might be no more than technology-intensive societies. An explanation of historical change that meets this standard of clarity is surely more useful than one in which the theory of change is hidden by less precise language.

Before considering how the three accounts of technological determinism measure up to this test, some clarification is needed for what the word "technology" is to mean here. The term is typically defined in several ways. Most restrictively, it is taken to mean "artifact." Other definitions are increasingly inclusive, incorporating processes for pro-

13. The concept of determinism has been interpreted in a variety of ways. Theological determinism sees human actions as deriving from hand of God in such a way that humans do not exercise discretion over their future. One account of psychological determinism sees the mind and human will as epiphenomenal, because psychological states are derived strictly from the physical laws governing the biological material in the brain. In economic determinism, self-interest and utility-seeking by people drive economic behavior, from which other social and political phenomena arise.

14. Heilbroner departs from his original account when he proposes a soft determinism in "Technological Determinism Revisited." By my definition, his new theory does not represent technological determinism.

ducing artifacts, knowledge about artifacts and process, and even systems of organization and control. These broader definitions may be useful elsewhere, but they become problematic when used in the context of a theory of technological determinism. Because factors such as knowledge and forms of social organization are important distinguishing features of societies, treating them as features of technology conflates causes and effects. If technological determinism is to provide a causal theory of how technology produces changes in society, then technology and society must be kept definitionally distinct.

Including factors broader than artifacts or machines in the definition of technology, and hence in the explanation of the determining factors for human history, requires the conclusion that social change is dependent in part upon social factors. However appealing that model is, it cannot meet the standard of being technological and deterministic, because it makes society at least partly responsible for its own future. The label "technological determinism" can only confuse such a theory. For these reasons I assume that, to make sense, technological determinism must define technology as physical artifact or machine and the associated material elements by which these are produced.

How do Normative accounts stand up to the requirement of being both technological and deterministic? Because these accounts attribute causal agency in the history of technology to human social practice and beliefs, rather than to technology or prior technological laws, they fail on both grounds. They do not impute special force to the characteristics of technology as artifacts, nor do they posit that the surrogate norm of efficiency be permanent or predetermined and out of reach of human control. On the contrary, the phenomena described by Normative Accounts are assumed to be the products of human action and intellect. It may vary across time and culture. Although these phenomena may be highly resistant to other forces, they are ultimately subject to the influence of all the forces that we typically think of as producing social change. Normative accounts actually deconstruct technological determinism into a cultural phenomenon or condition found in certain societies. This view is done a disservice by the confusing label "technological determinism."

Unintended Consequences accounts also fail as forms of technological determinism. As Winner points out, such accounts really amount to indeterminism.[15] That an outcome is unpredictable and uncontrollable may be significant, but this does not require that it be "determined."[16] Many human activities (e.g. the passage of law, the selection of a political leader, or the creation of a treaty between nations) reveal the properties of unpredictability or uncontrollability of final outcomes. Unintended consequences are basic facets of social action, rather than the special products of technology. Unintended Consequences accounts do not justify attributing the unpredictability of social outcomes to features of technology. These accounts are also neither technological nor deterministic.

If there is to be any meaningful interpretation of technological determinism, this leaves Nomological accounts as candidates. These accounts make the strongest claim about social change, a claim tied directly to technology. Indeed, they meet the criteria by holding that society evolves along a fixed and predetermined path, regardless of human intervention. That path is itself given by the incremental logic of technology and its parent, science. Here is a claim for the technological determination of history. Nomological accounts meet both tests.

What does the selection of only Nomological accounts as truly technologically deterministic mean for debates about technology in history? Most important, it encourages the choice of a different label for theories that I call Normative or Unintended Consequences accounts. In his essay in this volume, Thomas P. Hughes suggests the term "technological momentum" to describe the increasing capacity of technological systems to influence societies as those systems grow in size. Hughes sides with those who bring a "reciprocal relations" view to the interpretation of the history of technology. Both of these concepts—momentum and reciprocal causation—offer more precision and clarity to many theories of history than does "technological determinism." More judicious use of that label seems wise.

Applying This Model: Karl Marx as an Example

One test of the usefulness of thinking about technological determinism in terms of Nomological, Normative, and Unintended Conse-

15. Winner, *Autonomous Technology*, p. 88ff.

16. The observer of an unanticipated outcome cannot conclude whether the event was the result of unknown deterministic laws or whether the event was undetermined.

quences accounts is to bring this typology to the recurring debate about Marx. As I suggested above, those who have argued over whether Marx's historical materialism constitutes technological determinism seem to disagree as much about what technological determinism means as about anything else.[17] Winner argues that Marx established the first "coherent theory of autonomous technology."[18] Hansen claims that Marx conceived of social processes not in economic but in technological terms.[19] Mumford tells us that Marx assigned technology the "central place and directive function in human development."[20] Heilbroner connects the "basic Marxian paradigm" to technological determinism.[21]

Yet Rosenberg claims Marx saw historical change as a social rather than a technological process, especially in the case of the emergence of capitalist markets from feudalism without the influence of any major technological achievements.[22] MacKenzie claims Marx was not a technological determinist because the "forces of production," Marx's engine of history, do not equate to technology.[23] Miller helps Marx escape the label by arguing that work relations are an independent force in history.[24]

We can make sense of this debate by adhering to the requirements of Nomological accounts as a benchmark for technological determin-

17. Several of Marx's aphorisms are commonly cited as "proof" of what he thought about technology. From the preface to *A Contribution to the Critique of Political Economy:* "Men enter into definite relations that are indispensable and independent of their will, relations of production which correspond to a definite state of development of their material productive forces." From *The Poverty of Philosophy:* "The hand-mill gives you society with the feudal lord, the steam-mill, society with the industrial capitalist." For commentary on the context and use of these passages see Miller, *Analyzing Marx,* p. 174ff; Rürup, "Historians and Modern Technology," p. 180ff; Jon Elster, *Making Sense of Marx* (Cambridge University Press, 1985), pp. 270–271.

18. Winner, *Autonomous Technology,* p. 39.

19. Hansen, "The Technological Interpretation of History."

20. Mumford, *The Myth of the Machine,* p. 4. Mumford also argues the same point in the companion volume, *The Pentagon of Power.*

21. Heilbroner, "Do Machines Make History?" pp. 335–345.

22. Rosenberg, "Marx as a Student of Technology," pp. 56–57.

23. MacKenzie, "Marx and the Machine."

24. Miller, *Analyzing Marx,* p. 188ff.

ism. Recall the two components of this standard. If Marx's historical materialism is to stand up to this test, it must hold that

1. social change be causally determined by preceding phenomena or laws

and

2. these laws necessarily depend upon features of technology for their logic or as their vehicle.

These requirements focus the discussion of the role of technology in Marx's theory of history.[25]

An understanding of Marx's view of technology must be informed by an interpretation of his account of the forces of production. There is little doubt that, for Marx, whatever significance technology has derives from its relationship to economic activity—its role as a productive force—rather than from any other social or historical influence it might have. The connection between technology and productive forces is the central factor in the problem, and it is at this point that agreement about Marx and technological determinism has traditionally begun to unravel. Deciding whether historical materialism represents technological determinism requires asking whether Marx's views of the forces of production meet our two criteria.

Asking whether Marx's historical materialism represents technological determinism is related to asking whether it represents economic determinism, and this distinction is a good starting point for analysis. I understand "economic determinism" to mean that the tendencies, forces, and outcomes of economic processes exert an independent, determining influence on other aspects of social development, such as political organization and cultural beliefs. This notion, like technological determinism, holds that all aspects of society are dependent upon a single underlying set of factors for development and change. What distinguishes the two determinisms is that

25. When referring to Marx's "view" or "theory," I rely primarily upon *The German Ideology,* the *Grundrisse,* and *Capital* to construct a composite Marxian account, despite the fact that these works span several periods in Marx's work and occasionally contain statements that do not appear entirely consistent with one another in all their details. *The German Ideology* is useful here mainly for the discussion of human nature and its relationship to the natural order. The *Grundrisse* is helpful for its discussion of machinery and automation, and *Capital* for its analysis of capitalist production.

technological determinism limits those factors to the characteristics of technology, while economic determinism might include a number of factors (e.g. economic organization, natural resources, human capacities for productive work, the availability of potential trading partners).

Joshua Cohen and Robert Brenner have argued that political obstacles such as collective action problems can be sufficient to overcome or block inherent tendencies for productive development.[26] Theirs amounts to a nondeterministic interpretation of historical materialism. On this view, productive forces do not exercise primacy; they alone cannot determine the course of social development. Problems of incentive, class power, organization, or political resources can interrupt or reverse trends deriving from the economic determinism. If Marxian history cannot be attributed to economic factors, it cannot be attributed to technology.[27]

The more difficult question is: What if the productive forces are taken to be primary in the determination of social and political life, as G. A. Cohen has argued? Does an interpretation of historical materialism on which economic factors determine social change commit one to technological determinism in the form I have defined it?

G. A. Cohen's defense of Marx's theory of history is a good place from which to start to answer this question. Cohen's development Thesis and Primacy Thesis raise the fundamental issues for the present inquiry: How are the forces of production to be defined, and to what can their primacy over other factors and their tendency for development be attributed?[28] What is the relationship of technology to these forces?

26. Joshua Cohen, "Book Review, *Karl Marx's Theory of History: A Defense,* by G. A. Cohen," *Journal of Philosophy* 79 (1982): 253–273; Robert Brenner, "The Origins of Capitalist Development: A Critique of Neo-Smithian Marxism," *New Left Review* 104 (July–August 1977): 25–92.

27. This is true unless one introduces some other mechanism besides economic activity through which technology exerts its social influence. While such an approach to technological determinism is feasible, it is clearly not commensurate with Marx's view of history. If Marx is a technological determinist, he must be so through the involvement of technology in productive activity.

28. Cohen's two theses form the core of his defense of Marx's theory. In summary, the Development Thesis is that "the productive forces tend to develop throughout history: and the Primacy Thesis is that "the nature of

The first question asks how Marx defines the forces of production. Marx himself identifies three factors in the labor process: the activity of people, the subject of work, and the instruments of work.[29] Cohen categorizes these into two groups—means of production and labor power[30]—and I will follow his approach. Labor power consists of the human faculties that contribute to productive effort: strength, skill, knowledge, inventiveness, and so on. Means of production, on the other hand, include spaces, raw materials, and (most important here) technology and the instruments of production (which are, as Marx points out, themselves the products of previous production processes).[31]

A somewhat controversial question arises at this point: Should relations of production be included among the forces of production? Miller takes the position that they should, arguing against the view that Marx was a technological determinist on the basis that "work relations" are to be included in a broader definition of the productive forces. He construes much of the issue of technological determinism as a question about whether work relations are an independent productive force. But G. A. Cohen's dyad of means of production and labor power specifically excludes work relations for the following reason: They are the primary factor over which the forces of production are to exert their primacy; they are the dependent variable of explanation. Miller's approach to relations of production is therefore unhelpful. It is subject to the standard objection to Marxist social theory pointed out by Joshua Cohen: One cannot explain the state of the relations of production in terms of the forces of production if the former are to be subsumed under the latter.[32] For this reason I will distinguish the two.

the production relations of a society is explained by the level of development of its productive forces." (G. A. Cohen, *Karl Marx's Theory of History,* p. 134ff.)

29. Karl Marx, *Capital,* volume I, chapter VII, in *The Marx-Engels Reader,* ed. R. Tucker (Norton, 1978), p. 345. (All subsequent references to works by Marx refer to this volume.)

30. Cohen, *Karl Marx's Theory of History,* chapter II.

31. Marx also includes science as a productive force, and this is significant for the question of technological determinism. See *Grundrisse,* pp. 281–282. But science only enters the production process at a particular stage in history, through its practical application in technology, and so Cohen is correct in leaving it from his basic catalog.

32. Cohen, *Karl Marx's Theory of History,* p. 259.

What then is the role of technology as one of the means of production? The criteria for technological determinism would require that the development and primacy of the productive forces be dependent upon some internal logic of technological enterprise where, for example, telegraphs or the system of wage labor employed in steel mills are attributable to railroad technology rather than to collective or individual social choice. For historical materialism to be technologically deterministic, the overall growth in human labor and the availability of technology and other means of production must derive from the internal characteristics of technology itself.

But Marx simply does not connect productive power to technology in this way. His description of the growth of productive forces is best understood by dividing economic history into two phases and considering the role of technology in each.[33] The first phase covers the feudal, trade-oriented, and early manufacturing-oriented eras. Marx's account suggests that economic change in this phase is dependent primarily upon Smithian division of labor, where different forms of division of labor represent different forms of ownership and where the most significant of these divisions is that between town and country.[34] In this phase, the productive forces develop through the increasing division of labor. Technology is of little account in these changes. Here, Marx's well-known claim that how far the productive forces have developed is shown by the division of labor is most applicable.[35] From the feudal era through the early manufacturing era, technology is not a primary factor in social change. This point alone might disqualify Marx as a technological determinist, since true determinism can hardly explain some eras of history but not others.

Yet even in industrial periods Marx does not rely on technology as a primary causal agent, an unmoved mover. While technology does enter the picture more significantly in the second phase of historical growth (involving automation and the development of industrial capitalism), it is not the first cause of this growth. Marx describes technology's role in his well-known discussion of automation under

33. Presumably, if Marx's theory of history is truly technologically deterministic, technology can explain all phases of economic history.

34. Marx, *German Ideology,* pp. 150–155, 159–163.

35. Ibid., p. 150.

industrial capitalism in the *Grundrisse*.[36] The most important feature in the development of the forces of production in this period is the transformation of the means of production into fixed capital in the form of the machine. Automation transforms the machine from a means of transmitting human action to the object of labor into the actual doer of the work.

But the role of technology in this process is easily misunderstood. Marx's indication that highly developed machinery is required for this historical step should not be construed as a claim that technology is a cause or a source of it, or that its appearance is sufficient. Marx believed that during industrialization scientific knowledge, in the form of "mechanical laws," enters the production process as a productive force through its practical application in technology.[37] The incorporation of automation into production was a necessary step for historical development. Yet the development of machinery to this point depends upon prior social processes. In no way is this entrance of technology into the productive process independent of social history, as would be required by technological determinism. Marx says that it "occurs only when large industry has already reached a higher state and all the sciences have been pressed into the service of capital, and when available machinery already provides great capabilities."[38] The point is that industry and technology can reach this higher phase of automation and productivity only through the division of labor—the "transformation of workers' operations into more and more mechanical ones"—and through "the process of capital accumulation."[39] So, while the "productive forces of large industry depend upon science and technology," at this point productive forces themselves, science and technology "are in turn related to the development of material production."[40] The introduction of technology into the labor process is contingent upon social organization, specialization, and aggregation of wealth. In this second historical phase, Marx portrays technology more as an enabling factor than as an original cause, autonomous force, or determining agent.

36. Marx, *Grundrisse,* p. 279.

37. Ibid., pp. 281–283.

38. Ibid.

39. Ibid.

40. Ibid.

Further, the real significance of the eventual transformation of production into a technological process stems not from the fact that technology has become the doer of work but from the fact that control over individual productive activity has become further removed from the laborers—that they have become more fully "alienated" from the productive process. Technology functions as the device by which owners of the means of production further alienate the proletariat from their own work. This instrumental use of technology by the bourgeoisie for their own ends makes technology important in the capitalist phase of history. Technology itself neither causes not necessitates the class struggle that follows.

The important point here is that a theory of human nature underlies Marx's description of the role of automation in industrialization. Marx is describing a process that is dependent not upon features of technology but upon human characteristics, such as the drive to accumulate and the resistance to alienation. These compel the development of the forces of production and their influence over social and political life. This theory, which Marx adapted from Hegel, is his engine of history. Can technological determinism be built on a theory of human nature in this way? It cannot. For Marx, technology was no more than one kind of important and efficient fuel for history's human engine.

It is worth specifying just what kind of characteristics Marx assigned humans in his theory, in order to see their relationship to the availability of industrial technology. What follows is a synopsis of Marx's account of human nature in *The German Ideology* and *Capital.*

Factors Driving the Forces of Production

1. *A basic human drive for self-expression.* It is a fundamental characteristic of humans that they are driven to outwardly express their own being or "mode of life." This aspect of human nature will also lead people to resist any forces that tend to prevent self-expression or that alienate people from the means of that expression.

2. *Productive activity forms the fundamental mode of self-expression.* Production of the means of subsistence and the satisfaction of human needs is the basic method by which humans fulfil their drive to express themselves. This second characteristic derives from Marx's anti-Hegelian "first premise of history"—the existence of living human individuals—as well as from the related issue of human orga-

nization and relationship to nature. This productive activity forms the basis of the economy and provides for its social significance.

3. *The expansion of human needs*. The basic needs and wants for whose satisfaction commodities are produced expand throughout history. This necessitates continued and increasing production of use values, and, as Marx says, "the nature of wants, whether springing from the stomach or from fancy, doesn't matter."[41] This provides for a state of perpetual scarcity of a sort, and is more satisfactory than an explanation of scarcity as environmentally derived, since scarcity is so highly variable throughout history and place.

Where does this leave technology? Marx indicates that there are several historical or environmental conditions that enable and facilitate the productive growth deriving from the above three factors. Among these conditions are increasing population and expanding world markets for trade. These assist in the creation of the labor market and provide the basis for competition and the growth of social intercourse.[42] These conditions contribute, for example, to the phenomenon of the value of labor power being worth less than the use value that it creates.[43]

The presence of science and technology is also an enabling condition. Their availability as productive forces at historically significant point makes possible the accelerated accumulation of capital by the bourgeoisie. But the entrance of science and technology into the production process is contingent upon other factors, such as the increasing ownership of the means of production by capitalists. Certain new technologies facilitate the continuation of a process underway since long before their emergence.

Marx's view of the connection of human characteristics to technology in the development of the forces of production may be summarized in the following way. A set of human attributes provides the logic of the development of the forces of production and their primacy over other features of social and political life. Another set of environmental conditions facilitate development at various points in history. Technology is among them.

41. Marx, *Capital*, p. 303.

42. Marx, *German Ideology*, pp. 159–161.

43. Marx, *Capital*, p. 357.

Necessary Factors in Marx's Theory of History

1. a basic drive for self-expression
2. the form of self-expression is production
3. expanding needs

Conditions That Facilitate Productive Development in History

1. expanding population
2. increasing social intercourse
3. the availability of science and technology, especially in the later phase of capitalism

The relationship of technology to other forces of production and to the relations of production outlined here does not meet both of the criteria for technological determinism that are manifest in the Nomological account. While history does proceed according to a fixed, law-driven sequence, technological change is not the primary factor in social development—especially in feudal, trade-oriented, and early manufacturing eras. It is significant only in capitalist economies, and then as an enabling factor. In Marx's scheme, technology is pertinent only to a part of human history; it is not responsible for all of it. On Marx's view, the division of labor, the labor time, and the level of "alienation" inherent in a set of production relations are factors of more lasting importance than technology. When technology does enter this historical process in a central way, it is important only because of the manner in which it facilitates increases in the capital-accumulation process already taking place.

Marxian history cannot fulfil the requirements of a Nomological account of technological determinism. It clearly meets the first criterion (Marx was a determinist), but it fails the second (He was an economist more than a technologist).

Marx claimed that history is the development of the labor process into a social process, and this suggests what is perhaps the best way to view his conception of the role of technology. On his account, technology is used instrumentally by human actors whose actions are, in a collective sense, historically determined by their own characteristics. The intentional use of technology by human actors is an important theme in Marx's work, one quite contradictory in nature to real technological determinism. Consider Marx's claim that the inherent tendency of the introduction of automation in manufacturing, based upon the characteristics of the technology, would be a decrease in the length of the workday. Yet the actual result of automation, believed Marx, is the lengthening of the workday. This happens because capitalist owners of technology compensate for increased productivity in order to further increase total productive output.[44] In this case, whatever natural or inherent effects technology tends to produce are overcome by willful human actions.

The monumental increase in the forces of production which Marx foresaw in postcapitalistic society was to come not in the first instance from new technology, but because non-Smithian conditions would then prevail. The extraction of surplus labor would no longer be necessary, not because of new technology but because of a new self-awareness on the part of workers and because of new relations of production in which "labor has become not only means of life but life's prime want."[45] For Marx, the final achievement of economic development in history was to be a nontechnological one. Marx had in mind that technology is in the ultimate service of humanity, not the other way around.

Conclusion

If Marx was no technological determinist, then there may be few who are. The standards of the Nomological account of technological determinism are stringent, and it is hard to imagine a theory of history meeting this definition that would be plausible. Requiring our definition of technological determinism to be precise, semantically consistent, and meaningfully distinct from other explanations of history and social change leaves us with a rather unlikely account—one for which not even historical materialism seems to qualify. What is left, after Normative and Unintended Consequences accounts are stripped away (for these cannot truly qualify as forms of determinism), is a view of history in which human will has no real role—in which culture, social organization, and values derive from laws of nature that are manifest through technology.

The value of forcing technological determinism to retreat into the philosopher's corner of strict laws and the absence of free will may

44. Ibid., p. 406.

45. Marx, *A Critique of the Gotha Programme,* p. 531.

be that we are finally able to dispense with it as an intruder into discussions of the history of technology. Reserving this label for the austere requirements of Nomological accounts may allow us to speak of normative theories of history, unintended-consequences accounts, and so on without the confusing terminology of "technological determinism" interrupting. As interpretive tools, these concepts seem richer and more useful, and only muddled by language that invokes determination and laws of technology.

Acknowledgments

An earlier version of this essay appeared as "Karl Marx and the Three Faces of Technological Determinism" in *Social Studies of Science* (20 (1990): 331–351), and the present version is published with permission from Sage Publications, Ltd. I appreciate helpful comments from Joshua Cohen, Kenneth Keniston, Leo Marx, and Neil Schaefer.

Technological Momentum

Thomas P. Hughes

In 1969 Thomas Hughes published an article about hydrogenation in post-World War I Germany in which he coined the phrase "technological momentum" to explain how that nation's leading chemical manufacturer subsequently became linked with Adolf Hitler's Nazi regime. Ever since, historians have debated the relationship of technological momentum to technological determinism—particularly, whether they are synonymous in meaning and, if they are not, how they differ.

In the essay that follows, Hughes provides a useful clarification by locating the concept of technological momentum "somewhere between the poles of technological determinism and social constructivism." Through a series of examples drawn from his extensive research on the emergence of technological systems, he shows that younger developing systems tend to be more open to sociocultural influences while older, more mature systems prove to be more independent of outside influences and therefore more deterministic in nature. Hughes views technological momentum as an alternative to technological determinism and contends that it is a more valuable interpretative concept than either technological determinism or social constructivism because it is time dependent yet sensitive to the messy complexities of society and culture. To some critics, however, Hughes's systems-oriented explanation of the past remains essentially deterministic because it places technology at the center of the historical process and links everything else to it. Hughes naturally denies this charge, emphasizing that technological momentum is an integrative concept that gives equal weight to social and technical forces. Whether or not one agrees with Hughes's interpretative framework, his emphasis on the momentum of technological systems has helped to define more precisely the differences between the technological and social determinants of change.

The concepts of technological determinism and social construction provide agendas for fruitful discussion among historians, sociologists, and engineers interested in the nature of technology and technological change. Specialists can engage in a general discourse that subsumes their areas of specialization. In this essay I shall offer an additional concept—technological momentum—that will, I hope, enrich the discussion. Technological momentum offers an alternative to technological determinism and social construction. Those who in the past espoused a technological determinist approach to history offered a needed corrective to the conventional interpretation of history that virtually ignored the role of technology in effecting social change. Those who more recently advocated a social construction approach provided an invaluable corrective to an interpretation of history that encouraged a passive attitude toward an overwhelming technology. Yet both approaches suffer from a failure to encompass the complexity of technological change.

All three concepts present problems of definition. Technological determinism I define simply as the belief that technical forces determine social and cultural changes. Social construction presumes that social and cultural forces determine technical change. A more complex concept than determinism and social construction, technological momentum infers that social development shapes and is shaped by technology. Momentum also is time dependent. Because the focus of this essay is technological momentum, I shall define it in detail by resorting to examples.

"Technology" and "technical" also need working definitions. Proponents of technological determinism and of social construction often use "technology" in a narrow sense to include only physical artifacts and software. By contrast, I use "technical" in referring to physical artifacts and software. By "technology" I usually mean technological or sociotechnical systems, which I shall also define by examples.

Discourses about technological determinism and social construction usually refer to society, a concept exceedingly abstract. Historians are wary of defining society other than by example because they have found that twentieth-century societies seem quite different from twelfth-century ones and that societies differ not only over time but over space as well. Facing these ambiguities, I define the social as the world that is not technical, or that is not hardware or technical software. This world is made up of institutions, values, interest groups, social classes, and political and economic forces. As the reader will learn, I see the social and the technical as interacting within technological systems. Technological system, as I shall explain, includes both the technical and the social. I name the world outside of technological systems that shapes them or is shaped by them the "environment." Even though it may interact with the technological system, the environment is not a part of the system because it is not under the control of the system as are the system's interacting components.

In the course of this essay the reader will discover that I am no technological determinist. I cannot associate myself with such distinguished technological determinists as Karl Marx, Lynn White, and Jacques Ellul. Marx, in moments of simplification, argued that waterwheels ushered in manorialism and that steam engines gave birth to bourgeois factories and society. Lenin added that electrification was the bearer of socialism. White elegantly portrayed the stirrup as the prime mover in a train of cause and effect culminating in the establishment of feudalism. Ellul finds the human-made environment structured by technical systems, as determining in their effects as the natural environment of Charles Darwin. Ellul sees the human-made as steadily displacing the natural—the world becoming a system of artifacts, with humankind, not God, as the artificer.[1]

Nor can I agree entirely with the social constructivists. Wiebe Bijker and Trevor Pinch have made an influential case for social construction in their essay "The Social Construction of Facts and Artifacts."[2] They argue that social, or interest, groups define and give meaning to artifacts. In defining them, the social groups determine the designs of artifacts. They do this by selecting for survival the designs that solve the problems they want solved by the artifacts and that fulfill desires they want fulfilled by the artifacts. Bijker and Pinch emphasize the interpretive flexibility discernible in the evolution of artifacts: they believe that the various meanings given by social groups to, say, the bicycle result in a number of alternative designs of that machine.

1. Lynn White, Jr., *Medieval Technology and Social Change* (Clarendon, 1962); Jacques Ellul, *The Technological System* (Continuum, 1980); Karl Marx, *Capital: A Critique of Political Economy,* ed. F. Engels; *Electric Power Development in the U.S.S.R.,* ed. B. I. Weitz (Moscow: INRA, 1936).

2. The essay is found in *The Social Construction of Technological Systems: New Directions in the Sociology and History of Technology,* ed. W. E. Bijker et al. (MIT Press, 1987).

The various bicycle designs are not fixed; closure does not occur until social groups believe that the problems and desires they associate with the bicycle are solved or fulfilled.

In summary, I find the Bijker-Pinch interpretation tends toward social determinism, and I must reject it on these grounds. The concept of technological momentum avoids the extremism of both technological determinism and social construction by presenting a more complex, flexible, time-dependent, and persuasive explanation of technological change.

Technological Systems

Electric light and power systems provide an instructive example of technological systems. By 1920 they had taken on a messy complexity because of the heterogeneity of their components. In their diversity, their complexity, and their large scale, such mature technological systems resemble the megamachines that Lewis Mumford described in *The Pentagon of Power.*[3] The actor networks of Bruno Latour and Michel Callon[4] also share essential characteristics with technological systems. An electric power system consists of inanimate electrons and animate regulatory boards, both of which, as Latour and Callon suggest, can be intractable if not brought in line or into the actor network.

The Electric Bond and Share Company (EBASCO), an American electric utility holding company of the 1920s, provides an example of a mature technological system. Established in 1905 by the General Electric Company, EBASCO controlled through stock ownership a number of electric utility companies, and through them a number of technical subsystems—namely electric light and power networks, or grids.[5] EBASCO provided financial, management, and engineering

3. Lewis Mumford, *The Myth of the Machine: II. The Pentagon of Power* (Harcourt Brace Jovanovich, 1970).

4. Bruno Latour, *Science in Action: How to Follow Scientists and Engineers through Society* (Harvard University Press, 1987); Michel Callon, "Society in the Making: The Study of Technology as a Tool for Sociological Analysis," in *The Social Construction of Technological Systems.*

5. Before 1905, General Electric used the United Electric Securities Company to hold its utility securities and to fund its utility customers who purchased GE equipment. See Thomas P. Hughes, *Networks of Power: Electrification in Western Society, 1880–1930* (Johns Hopkins University Press, 1983), pp. 395–396.

construction services for the utility companies. The inventors, engineers, and managers who were the system builders of EBASCO saw to it that the services related synergistically. EBASCO management recommended construction that EBASCO engineering carried out and for which EBASCO arranged financing through sale of stocks or bonds. If the utilities lay in geographical proximity, then EBASCO often physically interconnected them through high-voltage power grids. The General Electric Company founded EBASCO and, while not owning a majority of stock in it, substantially influenced its policies. Through EBASCO General Electric learned of equipment needs in the utility industry and then provided them in accord with specifications defined by EBASCO for the various utilities with which it interacted. Because it interacted with EBASCO, General Electric was a part of the EBASCO system. Even though I have labeled this the EBASCO system, it is not clear that EBASCO solely controlled the system. Control of the complex systems seems to have resulted from a consensus among EBASCO, General Electric, and the utilities in the systems.

Other institutions can also be considered parts of the EBASCO system, but because the interconnections were loose rather than tight[6] these institutions are usually not recognized as such. I refer to the electrical engineering departments in engineering colleges, whose faculty and graduate students conducted research or consulted for EBASCO. I am also inclined to include a few of the various state regulatory authorities as parts of the EBASCO system, if their members were greatly influenced by it. If the regulatory authorities were free of this control, then they should be considered a part of the EBASCO environment, not of the system.

Because it had social institutions as components, the EBASCO system could be labeled a sociotechnical system. Since, however, the system had a technical (hardware and software) core, I prefer to name it a technological system, to distinguish it from social systems without technical cores. This privileging of the technical in a technological system is justified in part by the prominent roles played by engineers, scientists, workers, and technical-minded managers in solving the problems arising during the creation and early history of

6. The concept of loosely and tightly coupled components in systems is found in Charles Perrow's *Normal Accidents: Living with High Risk Technology* (Basic Books, 1984).

a system. As a system matures, a bureaucracy of managers and white-collar employees usually plays an increasingly prominent role in maintaining and expanding the system, so that it then becomes more social and less technical.

EBASCO as a Cause and an Effect

From the point of view of technological—better, technical—determinists, the determined is the world beyond the technical. Technical determinists considering EBASCO as a historical actor would focus on its technical core as a cause with many effects. Instead of seeing EBASCO as a technological system with interacting technical and social components, they would see the technical core as causing change in the social components of EBASCO and in society in general. Determinists would focus on the way in which EBASCO's generators, by energizing electric motors on individual production machines, made possible the reorganization of the factory floor in a manner commonly associated with Fordism. Such persons would see street, workplace, and home lighting changing working and leisure hours and affecting the nature of work and play. Determinists would also cite electrical appliances in the home as bringing less—and more—work for women,[7] and the layout of EBASCO's power lines as causing demographic changes. Electrical grids such as those presided over by EBASCO brought a new decentralized regionalism, which contrasted with the industrial, urban-centered society of the steam age.[8] One could extend the list of the effects of electrification enormously.

Yet, contrary to the view of the technological determinists, the social constructivists would find exogenous technical, economic, political, and geographical forces, as well as values, shaping with varying intensity the EBASCO system during its evolution. Social constructivists see the technical core of EBASCO as an effect rather than a cause. They could cite a number of instances of social construction. The spread of alternating (polyphase) current after 1900, for instance, greatly affected, even determined, the history of the early

7. Ruth Schwartz Cowan, "The 'Industrial Revolution' in the Home," *Technology and Culture* 17 (1976): 1–23.

8. Lewis Mumford, *The Culture of Cities* (Harcourt Brace Jovanovich, 1970), p. 378.

utilities that had used direct current, for these had to change their generators and related equipment to alternating current or fail in the face of competition. Not only did such external technical forces shape the technical core of the utilities; economic forces did so as well. With the rapid increase in the United States' population and the concentration of industry in cities, the price of real estate increased. Needing to expand their generating capacity, EBASCO and other electric utilities chose to build new turbine-driven power plants outside city centers and to transmit electricity by high-voltage lines back into the cities and throughout the area of supply. Small urban utilities became regional ones and then faced new political or regulatory forces as state governments took over jurisdiction from the cities. Regulations also caused technical changes. As the regional utilities of the EBASCO system expanded, they conformed to geographical realities as they sought cooling water, hydroelectric sites, and mine-mouth locations. Values, too, shaped the history of EBASCO. During the Great Depression, the Roosevelt administration singled out utility holding-company magnates for criticism, blaming the huge losses experienced by stock and bond holders on the irresponsible, even illegal, machinations of some of the holding companies. Partly as a result of this attack, the attitudes of the public toward large-scale private enterprise shifted so that it was relatively easy for the administration to push through Congress the Holding Company Act of 1935, which denied holding companies the right to incorporate utilities that were not physically contiguous.[9]

Gathering Technological Momentum

Neither the proponents of technical determinism nor those of social construction can alone comprehend the complexity of an evolving technological system such as EBASCO. On some occasions EBASCO was a cause; on others it was an effect. The system both shaped and was shaped by society. Furthermore, EBASCO's shaping society is not an example of purely technical determinism, for EBASCO, as we have observed, contained social components. Similarly, social constructivists must acknowledge that social forces in the environment

9. More on EBASCO's history can be found on pp. 392–399 of *Networks of Power.*

were not shaping simply a technical system, but a technological system, including—as systems invariably do—social components.

The interaction of technological systems and society is not symmetrical over time. Evolving technological systems are time dependent. As the EBASCO system became larger and more complex, thereby gathering momentum, the system became less shaped by and more the shaper of its environment. By the 1920s the EBASCO system rivaled a large railroad company in its level of capital investment, in its number of customers, and in its influence upon local, state, and federal governments. Hosts of electrical engineers, their professional organizations, and the engineering schools that trained them were committed by economic interests and their special knowledge and skills to the maintenance and growth of the EBASCO system. Countless industries and communities interacted with EBASCO utilities because of shared economic interests. These various human and institutional components added substantial momentum to the EBASCO system. Only a historical event of large proportions could deflect or break the momentum of an EBASCO, the Great Depression being a case in point.

Characteristics of Momentum

Other technological systems reveal further characteristics of technological momentum, such as acquired skill and knowledge, special-purpose machines and processes, enormous physical structures, and organizational bureaucracy. During the late nineteenth century, for instance, mainline railroad engineers in the United States transferred their acquired skill and knowledge to the field of intra-urban transit. Institutions with specific characteristics also contributed to this momentum. Professors in the recently founded engineering schools and engineers who had designed and built the railroads organized and rationalized the experience that had been gathered in preparing roadbeds, laying tracks, building bridges, and digging tunnels for mainline railroads earlier in the century. This engineering science found a place in engineering texts and in the curricula of the engineering schools, thus informing a new generation of engineers who would seek new applications for it.

Late in the nineteenth century, when street congestion in rapidly expanding industrial and commercial cities such as Chicago, Baltimore, New York, and Boston threatened to choke the flow of traffic, extensive subway and elevated railway building began as an antidote. The skill and the knowledge formerly expended on railroad bridges were now applied to elevated railway structures; the know-how once invested in tunnels now found application in subways. A remarkably active period of intra-urban transport construction began about the time when the building of mainline railways reached a plateau, thus facilitating the movement of know-how from one field to the other. Many of the engineers who played leading roles in intra-urban transit between 1890 and 1910 had been mainline railroad builders.[10]

The role of the physical plant in the buildup of technological momentum is revealed in the interwar history of the Badische Anilin und Soda Fabrik (BASF), one of Germany's leading chemical manufacturers and a member of the I.G. Farben group. During World War I, BASF rapidly developed large-scale production facilities to utilize the recently introduced Haber-Bosch technique of nitrogen fixation. It produced the nitrogen compounds for fertilizers and explosives so desperately needed by a blockaded Germany. The high-technology process involved the use of high-temperature, high-pressure, complex catalytic action. Engineers had to design and manufacture extremely costly and complex instrumentation and apparatus. When the blockade and the war were over, the market demand for synthetic nitrogen compounds did not match the large capacity of the high-technology plants built by BASF and other companies during the war. Numerous engineers, scientists, and skilled craftsmen who had designed, constructed, and operated these plants found their research and development knowledge and their construction skills underutilized. Carl Bosch, chairman of the managing board of BASF and one of the inventors of the Haber-Bosch process, had a personal and professional interest in further development and application of high-temperature, high-pressure, catalytic processes. He and other managers, scientists, and engineers at BASF sought additional ways of using the plant and the knowledge created during the war years. They first introduced a high-temperature, high-pressure catalytic process for manufacturing synthetic methanol in the early 1920s. The momentum of the now-generalized process next showed itself in management's decision in the mid 1920s to invest in research and development aimed at using high-temperature, high-pressure

10. Thomas Parke Hughes, "A Technological Frontier: The Railway," in *The Railroad and the Space Program,* ed. B. Mazlish (MIT Press, 1965).

catalytic chemistry for the production of synthetic gasoline from coal. This project became the largest investment in research and development by BASF during the Weimar era. When the National Socialists took power, the government contracted for large amounts of the synthetic product. Momentum swept BASF and I.G. Farben into the Nazi system of economic autarky.[11]

When managers pursue economies of scope, they are taking into account the momentum embodied in large physical structures. Muscle Shoals Dam, an artifact of considerable size, offers another example of this aspect of technological momentum. As the loss of merchant ships to submarines accelerated during World War I, the United States also attempted to increase its indigenous supply of nitrogen compounds. Having selected a process requiring copious amounts of electricity, the government had to construct a hydroelectric dam and power station. This was located at Muscle Shoals, Alabama, on the Tennessee River. Before the nitrogen-fixation facilities being built near the dam were completed, the war ended. As in Germany, the supply of synthetic nitrogen compounds then exceeded the demand. The U.S. government was left not only with process facilities but also with a very large dam and power plant.

Muscle Shoals Dam (later named Wilson Dam), like the engineers and managers we have considered, became a solution looking for a problem. How should the power from the dam be used? A number of technological enthusiasts and planners envisioned the dam as the first of a series of hydroelectric projects along the Tennessee River and its tributaries. The poverty of the region spurred them on in an era when electrification was seen as a prime mover of economic development. The problem looking for a solution attracted the attention of an experienced problem solver, Henry Ford, who proposed that an industrial complex based on hydroelectric power be located along 75 miles of the waterway that included the Muscle Shoals site. An alliance of public power and private interests with their own plans for the region frustrated his plan. In 1933, however, Muscle Shoals became the original component in a hydroelectric, flood-control, soil-reclamation, and regional development project of enormous scope sponsored by Senator George Norris and the Roosevelt administration and presided over by the Tennessee Valley Authority. The tech-

11. Thomas Parke Hughes, "Technological Momentum: Hydrogenation in Germany 1900–1933," *Past and Present* (August 1969): 106–132.

nological momentum of the Muscle Shoals Dam had carried over from World War I to the New Deal. This durable artifact acted over time like a magnetic field, attracting plans and projects suited to its characteristics. Systems of artifacts are not neutral forces; they tend to shape the environment in particular ways.[12]

Using Momentum

System builders today are aware that technological momentum—or whatever they may call it—provides the durability and the propensity for growth that were associated more commonly in the past with the spread of bureaucracy. Immediately after World War II, General Leslie Groves displayed his system-building instincts and his awareness of the critical importance of technological momentum as a means of ensuring the survival of the system for the production of atomic weapons embodied in the wartime Manhattan Project. Between 1945 and 1947, when others were anticipating disarmament, Groves expanded the gaseous-diffusion facilities for separating fissionable uranium at Oak Ridge, Tennessee; persuaded the General Electric Company to operate the reactors for producing plutonium at Hanford, Washington; funded the new Knolls Atomic Power Laboratory at Schenectady, New York; established the Argonne and Brookhaven National Laboratories for fundamental research in nuclear science; and provided research funds for a number of universities. Under his guiding hand, a large-scale production system with great momentum took on new life in peacetime. Some of the leading scientists of the wartime project had confidently expected production to end after the making of a few bombs and the coming of peace.[13]

More recently, proponents of the Strategic Defense Initiative (SDI), organized by the Reagan administration in 1983, have made use of momentum. The political and economic interests and the organizational bureaucracy vested in this system were substantial—as its makers intended. Many of the same industrial contractors, research universities, national laboratories, and government agencies that took part in the construction of intercontinental ballistic missile systems,

12. On Muscle Shoals and the TVA, see Preston J. Hubbard's *Origins of the TVA: The Muscle Shoals Controversy, 1920–1932* (Norton, 1961).

13. Richard G. Hewlett and Oscar E. Anderson, Jr., *The New World, 1939–1946* (Pennsylvania State University Press, 1962), pp. 624–638.

National Air and Space Administration projects, and atomic weapon systems have been deeply involved in SDI. The names are familiar: Lockheed, General Motors, Boeing, TRW, McDonnell Douglas, General Electric, Rockwell, Teledyn, MIT, Stanford, the University of California's Lawrence Livermore Laboratory, Los Alamos, Hanford, Brookhaven, Argonne, Oak Ridge, NASA, the U.S. Air Force, the U.S. Navy, the CIA, the U.S. Army, and others. Political interests reinforced the institutional momentum. A number of congressmen represent districts that receive SDI contracts, and lobbyists speak for various institutions drawn into the SDI network.[14] Only the demise of the Soviet Union as a military threat allowed counter forces to build up sufficient momentum to blunt the cutting edge of SDI.

Conclusion

A technological system can be both a cause and an effect; it can shape or be shaped by society. As they grow larger and more complex, systems tend to be more shaping of society and less shaped by it. Therefore, the momentum of technological systems is a concept that can be located somewhere between the poles of technical determinism and social constructivism. The social constructivists have a key to understanding the behavior of young systems; technical determinists come into their own with the mature ones. Technological momentum, however, provides a more flexible mode of interpretation and one that is in accord with the history of large systems.

What does this interpretation of the history of technological systems offer to those who design and manage systems or to the public that might wish to shape them through a democratic process? It suggests that shaping is easiest before the system has acquired political, economic, and value components. It also follows that a system with great technological momentum can be made to change direction if a variety of its components are subjected to the forces of change.

For instance, the changeover since 1970 by U.S. automobile manufacturers from large to more compact automobiles and to more fuel-efficient and less polluting ones came about as a result of pressure brought on a number of components in the huge automobile

14. Charlene Mires, "The Strategic Defense Initiative" (unpublished essay, History and Sociology of Science Department, University of Pennsylvania, 1990).

production and use system. As a result of the oil embargo of 1973 and the rise of gasoline prices, American consumers turned to imported compact automobiles; this, in turn, brought competitive economic pressure to bear on the Detroit manufacturers. Environmentalists helped persuade the public to support, and politicians to enact, legislation that promoted both anti-pollution technology and gas-mileage standards formerly opposed by American manufacturers. Engineers and designers responded with technical inventions and developments.

On the other hand, the technological momentum of the system of automobile production and use can be observed in recent reactions against major environmental initiatives in the Los Angeles region. The host of institutions and persons dependent politically, economically, and ideologically on the system (including gasoline refiners, automobile manufacturers, trade unions, manufacturers of appliances and small equipment using internal-combustion engines, and devotees of unrestricted automobile usage) rallied to frustrate change.

Because social and technical components interact so thoroughly in technological systems and because the inertia of these systems is so large, they bring to mind the iron-cage metaphor that Max Weber used in describing the organizational bureaucracies that proliferated at the beginning of the twentieth century.[15] Technological systems, however, are bureaucracies reinforced by technical, or physical, infrastructures which give them even greater rigidity and mass than the social bureaucracies that were the subject of Weber's attention. Nevertheless, we must remind ourselves that technological momentum, like physical momentum, is not irresistible.

15. Max Weber, *The Protestant Ethic and the Spirit of Capitalism*, tr. T. Parsons (Unwin-Hyman, 1990), p. 155.

Retrieving Sociotechnical Change from Technological Determinism

Thomas J. Misa

Thomas Misa maintains that the level at which scholars conduct their research directly influences the degree of emphasis they place on technological change in their historical interpretations. Hence those who conduct "macro-level" studies are prone to technological determinism, while those who conduct "micro-level" studies are apt to find more contingent and multiple societal forces at work in the historical process. After discussing the characteristics of both approaches, Misa suggests a "middle-level" methodology that directs attention to the actors, institutions, and processes intermediate between the macro and the micro. He illustrates his point by looking at the origins of vertical integration at the Carnegie Steel company during the closing decades of the nineteenth century. As he shows, neither the "demands" of technology nor high-level strategy had much to do with the advent of vertical integration. Indeed, Andrew Carnegie and his senior partners seemed almost oblivious to its advantages. Only after a middle-level *official repeatedly pleaded with Carnegie to take action did he move toward the acquisition of invaluable iron ore properties in the upper midwest, and then almost as an afterthought. For Misa, the advantage of middle-range analysis is that it allows historians to combine the contingent, social construction of the micro approach with the "big picture" impact of the macro approach. Specifically it means focusing on institutions "intermediate" between the firm and the market and between the individual and the state. Like Thomas Hughes, Misa presses for an integrative approach to the past that will help historians better understand "the elaborate sociotechnical networks that span society."*

No theme is more vital to technology's storytellers, its practitioners, and citizens than that of technology and social change. While in recent years we have seen an explosion of empirical studies of technology, we have not seen a corresponding advance in our systematic analysis of this theme. Rarely do articles from the 1970s on other well-developed themes such as industrial research, technological systems, science-technology interactions, or engineering design read as well today as Reinhard Rürup's discussion of technological determinism.[1] In the last two decades, historians of technology have contributed impressively to our understanding of how society shapes technology but have nearly abandoned the urgent task of understanding how technology concurrently shapes society.[2] Those who have offered theories of technological change have deemphasized technology's role in social change.[3] An inadvertent result has been the message that technologies are well adapted to the societies and cultures that nurture them. Whatever the merits of the professional

1. Reinhard Rürup, "Historians and Modern Technology: Reflections on the Development and Current Problems of the History of Technology," *Technology and Culture* 15 (April 1974): 161–193. One example: Rürup observed (p. 162) that "modern technology has increasingly proved to be a system-stabilizing force, not merely one capable of revolutionizing social conditions"; this is a major theme of David F. Noble (*America by Design: Science, Technology, and the Rise of Corporate Capitalism* (Knopf, 1977)) as well as of Thomas P. Hughes (*American Genesis: A Century of Invention and Technological Enthusiasm, 1870–1970* (Viking Penguin, 1989)). Many of Rürup's themes—technology as ideology (*vide* Habermas), the political consequences of writing technological history as engineering history, the utility of Marxist concepts (productive forces, use value and exchange value) as well as traditional economic concepts in writing technological history, the possibility of controlling and steering the process of "technological progress"—have, for the most part, not been developed by historians of technology. On the other hand, several of Rürup's themes—the sociology of technology, industrial labor history, technological momentum—have been developed, in the pages of *Technology and Culture* and elsewhere.

2. John M. Staudenmaier, *Technology's Storytellers: Reweaving the Human Fabric* (MIT Press, 1985), 148–149, 198–199; Thomas J. Misa, "How Machines Make History, and How Historians (and Others) Help Them to Do So," *Science, Technology and Human Values* 13 (1988): 317–319.

3. Thomas J. Misa, "Theories of Technological Change: Parameters and Purposes," *Science, Technology and Human Values* 17 (1992): 3–12; Alex Roland, "Theories and Models of Technological Change: Semantics and Substance," *Science, Technology and Human Values* 17 (1992): 79–100.

strategy of consolidating the field around a set of well-defined core questions, the broader liabilities of this strategy have been unappreciated and have gone unremarked. Citizens facing the daunting sociotechnical problems of too much toxic waste and not enough ozone, among a dozen other challenges, may not appreciate accounts that point to the harmony of technology and culture. "We cannot afford," as William McNeill reminds us, "to make the world in which our fellow citizens live historically unintelligible."[4]

The time has come for technology's analysts to turn their collective attention to technology's society-shaping character. The time has come, too, to move beyond anecdotes, however useful for pedagogy the story of Robert Moses and Long Island's bus-blocking bridges may be. Toward this end, I will develop here a middle-level theory of technological determinism. I will argue that machines make history when historians and other analysts adopt a "macro" perspective, whereas a causal role for the machine is not present and is not possible for analysts who adopt a "micro" perspective. I will briefly discuss three historical questions in which the divide between the macro and the micro has strongly influenced our understanding of technology and social change and then will illustrate the middle-level theory by appraising, from both technological-determinist and social-determinist perspectives, a key structural change in mass-production steelmaking that inevitably features in the "master narrative" of American capitalism's development. I will conclude by proposing a methodological advance toward synthesizing the social-shaping-of-technology thesis with the technological-shaping-of-society antithesis.[5]

Such a synthesis requires more than the collecting of the right "facts"; it requires a reconceptualizing of the problem of technology and social change. A review of recent literature has suggested two patterns that explain when and how machines are seen as making history. First, disciplinary levels of analysis correlate strongly with disciplinary traditions of affirming or denying the thesis of tech-

4. William H. McNeill, *Mythistory and Other Essays* (University of Chicago Press, 1986), p. 95.

5. The essays in this volume by Phil Scranton, John Staudenmaier, and especially Thomas Hughes can be read as striving to achieve a similar goal. Scranton aims to dethrone the "master narrative" by broadening the field of inquiry.

nological determinism. Philosophers of technology adopting a macro-level analysis often affirm the thesis, whereas labor historians adopting a micro-level analysis typically deny the thesis, while business, urban, physical-science, and technological historians usually take intermediate positions. Conversely, across disciplines, authors affirming some version of technological determinism tend to adopt a macro perspective, while those denying technological determinism tend to adopt a micro perspective. This pattern—more than the author's politics or some other alleged social failing—explains how and where machines are permitted to make history.[6] Indeed, technological determinism may be a special case of the larger historical problem of continuity vs. change; at least, many general historians who desire to refute interpretations stressing fundamental change and support interpretations stressing continuity appear to rely on micro methods, too.[7]

6. Misa, "How Machines Make History," 308–331; see also Rosalind Williams's and Robert Heilbroner's essays in this volume. Contrast Michael Smith's social-critical essay in this volume, which implies that technological determinism is a species of wrong-thinking. My emphasis of this pattern is not to suggest that technological determinism does not have political *consequences* (only that manifest political *motives* need not be present for an author to adopt technological determinism); see Leo Marx, "American Literary Culture and the Fatalistic View of Technology," in *The Pilot and the Passenger: Essays on Literature, Technology, and Culture in the United States* (Oxford University Press, 1988), esp. p. 207: "The idea of technology as the controlling agent of our destiny lends itself to such [pastoral-romantic] retreats from politics. To invest a disembodied entity like 'The Machine' or 'technology' with the power to determine events is a useful way to justify disengagement from the public realm and a reversion to inaction and privacy. In the world of power as in the world of art, the pastoral response to change makes possible a consoling absolution from the painful complexity of political choice."

7. See such diverse micro-oriented authors as Alan Macfarlane, Christine Leigh Heyrman, and Michael Sherry. Macfarlane, in *The Culture of Capitalism* (Blackwell, 1987), stresses the continuity in social structure from medieval to early modern England. Heyrman, in *Commerce and Culture: The Maritime Communities of Colonial Massachusetts, 1690–1750* (Norton, 1984), argues against the "communal-breakdown" model of early New England social-economic development. Sherry, in *The Rise of American Air Power: The Creation of Armageddon* (Yale University Press, 1987), stresses the continuity in strategic thinking about air war from the early twentieth century to the atomic bomb. In contrast, William H. McNeill, in *The Human Condition: An Ecological and Historical View* (Princeton University Press, 1980), focuses on macro-social

In distinguishing between the macro and the micro, it is essential to emphasize that the issue is not merely the *size* of the unit of analysis.[8] Besides taking a larger unit of analysis, macro studies tend to *abstract* from individual cases, to impute *rationality* on actors' behalfs or posit *functionality* for their actions, and to be *order-driven*. Accounts focusing on these "order-bestowing principles" lead toward technological, economic, or ecological determinism. Conversely, accounts focusing on historical contingency and variety of experience lead away from all determinisms.[9] Besides taking a smaller unit of analysis, such micro studies tend to focus solely on case studies, to refute rationality or confute functionality, and to be disorder-respecting. Generally, macro studies make it easy for historical actors to appear rational, purposeful, and as key agents of change, whereas micro studies make it difficult or impossible for historical actors to have these same attributes. Consequently, since macro- and micro-oriented historians use significantly different concepts and language, their different methodologies have bedeviled attempts to test large interpretive schemes by means of detailed case studies—a vexing consideration that I will return to in the conclusion.

It is also helpful to disentangle the two subtheses most characteristic of technological determinism. Conveniently, exemplars of each are provided in James Beniger's book *The Control Revolution,*[10] a macro-

processes to give a causal account of 5000 years of human history inside 75 pages; but note his *micro*-level attack on the "Industrial Revolution" and other technologically defined periods of history (p. 5). As McNeill relates it, technological periodizations of history "got underway in the 1880s when Arnold Toynbee invented the Industrial Revolution. It neatly coincided with the reign of King George III (1760–1820) for the simple reason that Toynbee had been hired at Oxford to teach a course about the history of that reign."

8. I am indebted to Peter Perdue and Annemarie Mol for pointing out to me that such a two-pole analysis as this one does not permit one to plot the universe of social-historical analysis; nor do I intend it to. For instance, studies with a large scale but an interpretive methodology—for example, the history of *mentalités,* or Foucault-inspired studies generally—are not reducible to the macro-micro dimension. For a four-pole analysis that does aspire to plot the universe of social-historical analysis, see Charles Tilly, "How—and What—Are Historians Doing?" (New School for Social Research, Center for Studies of Social Change, Working Paper 58, January 1988).

9. On this point, contrast the essays in this volume by Heilbroner and Perdue with those of Scranton and Staudenmaier.

10. James R. Beniger, *The Control Revolution: Technological and Economic Ori-*

level technological-determinist account of the origins of the information age. The first subthesis asserts that technological change causes social change:

> Because technology defines the limits on what a society *can* do, technological innovation might be expected to be a major impetus to social change in the Control Revolution no less than in the earlier society transformations accorded the status of revolutions. [Indeed,] the Control Revolution resulted from innovation at a most fundamental level of technology—that of information processing.

The second asserts that technology is autonomous or independent of social influences:

> Each new technological innovation extends the processes that sustain life, thereby increasing the need for control and hence for improved control technology. This explains why technology appears autonomously to beget technology in general. . . .

Beniger exemplifies the tendency of macro analysts to abstract from individual cases (his case studies "prove" that the industrial system is transmogrified into a quasi-biological entity), to impute rationality to actors' behalfs (his actors perceive the "crisis of control" and solve it), and to be order-driven (his control revolution subsumes all other sociotechnical processes).[11]

gins of the Information Society (Harvard University Press, 1986), pp. 9–10. Beniger maintains that the Neolithic Revolution resulted from stone tools and plant and animal domestication, that the Commercial Revolution resulted directly from technical improvements in seafaring and navigation, and that the Industrial Revolution began with coal and steam power and a spate of new machinery for cotton textile manufacture.

11. Conversely, a micro-level interpretation upends Beniger's thesis. Of the "crisis of control" in production, Beniger argues (pp. 238–239) that it was caused by the increasing speed and volume of throughput: "Maintaining control of these accelerating throughputs proved to be a continuing struggle." "Solutions to this aspect of the control crisis . . . centered on two general information technologies, formal organization and preprocessing. The essential insight: to construct steel works as processors *explicitly*, with structures and procedures so determinate that information extraneous to any particular function would be eliminated on the designer's table, so to speak, and hence pose no further need for control." Beniger's case that "proves" his thesis is Carnegie Steel's Edgar Thomson works, a model of rational design for efficient throughput (*vide* Chandler). But Beniger's focus on ever-tightening systems of control blinds him to one of the most significant developments in

The micro/macro divide has worked, nearly invisibly, to skew historians' interpretations of many significant periods. Three examples illustrate the point. Compare two accounts of technology and social change during the eighteenth century. David Landes's *Unbound Prometheus* portrays a juggernaut: "The Industrial Revolution began in England in the eighteenth century, spread therefrom in unequal fashion to the countries of Continental Europe and a few areas overseas, and transformed in the span of scarce two lifetimes the life of Western man, the nature of his society, and his relationship to the other peoples of the world." Landes describes a macro-social process with technological change at its heart. The primary changes were the substitution of mechanical devices for human skills, the substitution of inanimate power (especially steam) for human and animal power, and the improvements in the getting and working of raw materials; concomitant changes occurred in industrial organization and in social and factory discipline. Abstracting from individual cases, imputing rationality, being order-driven, and picturing a technology-driven autonomous process—all these attributes are here: "In all of this diversity of technological improvement, the unity of the movement is apparent: change begat change."[12] Now consider Maxine Berg's *Age of Manufactures,* a compelling micro-level critique of Landes and other macro-level analysts. To comprehend sociotechnical change in eighteenth-century England, Berg looks closely at a wide range of industries (e.g., wool, worsted, silk, linen, and cotton) and work orga-

continuous-flow steelmaking: the Jones mixer, developed in the late 1880s, which decoupled iron smelting from steelmaking (see chapter 1 of my forthcoming book *A Nation of Steel: Steel and the Making of Modern America, 1865–1925.*

Beniger's assertion (p. 240) that "as late as the 1880s only a handful of managers and a small staff kept the E.T. running"—namely, Carnegie himself, three engineers, and one chemist—is simply wrong; in 1888, for instance, approximately 1000 workmen had something to do with keeping the E.T. running, according to Joseph Frazier Wall (*Andrew Carnegie* (Oxford University Press, 1970; University of Pittsburgh Press, 1989), p. 528).

12. David S. Landes, *The Unbound Prometheus: Technological Change and Industrial Development in Western Europe from 1750 to the Present* (Cambridge University Press, 1969), pp. 1–2, compare pp. 84–88 (disequilibrium in spinning and weaving). For a useful discussion of the assumptions underlying this and other conceptualizations, see Friedrich Rapp, "Structural Models in Historical Writing: The Determinants of Technological Development during the Industrial Revolution," *History and Theory* 21 (1982): 327–346.

nizations (e.g., household production, independent craftsmen, centralized workshops, and sweated labor, along with cottage and factory). She is able to explain where innovation occurred (for woolens and worsteds, in Yorkshire and Lancashire) and where it did not (the traditional textile centers of Essex, Norfolk, the West Country) by contrasting these regions' preexisting social structures, land-tenure patterns, and market orientations. For Berg, machines do not make history. She refutes the argument that Richard Arkwright's water-frame spinning machine required the rise of the factory system: "Arkwright's water frame . . . was originally designed as a small machine turned by hand and capable of being used in the home. It was Arkwright's [1769] patent which enclosed the machine within a factory, had it built only to large-scale specifications, and henceforth refused the use of it to anyone without a thousand-spindle mill."[13] Legal control of the water frame was feasible only if patent licenses were restricted to factory-scale units of production (as opposed to domestic-scale units, where unauthorized duplication would be difficult to monitor); in turn, factory-scale units were economical only when erected in a water-powered mill. Arkwright's fantastical profits cemented the popular view that his machinery and the factory system were one. This outcome should not obscure that an alternative path—domestic-scale water-frame spinning—was foreclosed not by technical necessity but by legal considerations imposed by Arkwright and his business partners. Focusing on case studies, unwinding technical rationality, and respecting the diversity of region- and sector-specific experiences—all these attributes are evident in Berg's "case for a more long-run, varied and complicated picture of the British path to Industrial Revolution."

We can now also shed new light on the old question: Was Karl Marx a technological determinist? The answer is no, and yes. Some commentators have been misled by the assumption that for Marx "productive forces" were somehow equivalent with technology. If they were equivalent, and if one adopts the primacy thesis, which

13. Maxine Berg, *The Age of Manufactures: Industry, Innovation and Work in Britain 1700–1820* (Oxford University Press, 1986), p. 243. Berg repeats the common technological-determinist claim that the flying shuttle doubled weavers' productivity (pp. 235, 246); the evidence is reviewed and the claim denied by Ako Paulinyi ("John Kay's Flying Shuttle: Some Considerations on His Technical Capacity and Economic Impact," *Textile History* 17 (1986): 149–166).

insists on the explanatory primacy of productive forces, then it follows that Marx must have been a technological determinist.[14] But technology is rarely defined so broadly as the forces of production. A more persuasive tactic here, deployed by Robert Heilbroner, Langdon Winner, and others, focuses on Marx's determinist aphorisms. Commentators wishing to portray Marx as a technological determinist have rarely failed to quote the famous passage from *The Poverty of Philosophy:* "The handmill gives you society with the feudal lord; the steam-mill, society with the industrial capitalist."[15] But, as Donald MacKenzie points out, not only is this aphorism historically inaccurate; it is also misleading. MacKenzie maintains that it is "extremely difficult to sustain" the thesis that Marx was a technological determinist "without invoking a peculiar and marked inconsistency between his general statements and particular analyses."[16] Similarly, Paul Adler points out that Marx varied the relative causal weight of technology depending on the time frame of his analysis.[17] Put another way: Marx the macro analyst penned statements (such as the above) that affirm technological determinism, but Marx the micro analyst rejected technological determinism.

Attempts to portray Marx as a productive-force determinist, as distinct from a technological determinist, have been reinvigorated with the rise of analytic Marxism in contemporary philosophy. The key work here is G. A. Cohen's *Karl Marx's Theory of History* (1978). Cohen attempts a serious, analytic philosophical defense of an "old-fashioned," "technological" interpretation of historical materialism, taking as his point of departure the 1859 preface to Marx's *Critique of Political Economy:*

In the social production of their life, men enter into definite relations that are indispensable and independent of their will, relations of production

14. See Misa, "How Machines Make History," pp. 311–312, and Bruce Bimber, "Karl Marx and the Three Faces of Technological Determinism," *Social Studies of Science* 20 (1990): 333–351.

15. Here Marx may have been following the lead of Engels, according to John M. Sherwood, "Engels, Marx, Malthus and the Machine," *American Historical Review* 90 (1985): 837–865.

16. Donald MacKenzie, "Marx and the Machine," *Technology and Culture* 25 (1984): 473–502, quote 480.

17. Paul S. Adler, "Marx, Machines, and Skill," *Technology and Culture* 31 (1990): 780–812, esp. 789.

which correspond to a definite state of development of their material productive forces. The sum total of these relations of production constitutes the economic structure of society, the real basis, on which rises a legal and political superstructure, and to which correspond definite forms of social consciousness. The mode of production of material life conditions the social, political and intellectual life process in general. It is not the consciousness of men that determines their being, but, on the contrary, their social being that determines their consciousness.

Cohen carefully distinguishes the forces of production (or productive forces) from the relations of production (or production relations). The forces of production comprise the means of production (that is, the instruments of production plus the raw materials) along with labor power (that is, the productive faculties of producing agents—strength, skill, knowledge, inventiveness, etc.); the relations of production alone constitute the economic structure. Cohen insists that for Marx the forces of production have explanatory primacy over the relations of production. Yet Cohen's is not simply a causal argument; it is a functional position:

. . . to say that an economic structure *corresponds* to the achieved level of the productive forces means: the structure provides maximum scope for the fruitful use and development of the forces, and obtains *because* it provides such scope. To say that being *determines* consciousness means, at least in large part: the character of the leading ideas of a society is explained by their propensity, in virtue of that character, to sustain the structure of the economic roles called for by the productive forces.[18]

Cohen's philosophical analysis, the essence of a macro-level interpretation, places him in stark contrast to other Marxist-inspired but micro-oriented authors—Harry Braverman, David Noble, and a generation of labor historians—who have emphasized class struggle by potentially revolutionary workers as a key motivation for capitalists to introduce new technologies.[19] In Marxism, in the Industrial Rev-

18. G. A. Cohen, *Karl Marx's Theory of History: A Defense* (Princeton University Press, 1978), quote vii, 28–32, 134–74, 180, quote 278–79. On analytic Marxism, see Alex Callinicos, ed., *Marxist Theory* (Oxford University Press, 1989), esp. 1–16, 88–104. Another author who generally affirms Marx's technological determinism, but qualifies it as "productive force" determinism is William H. Shaw; see "'The Handmill Gives You the Feudal Lord': Marx's Technological Determinism," *History and Theory* 18 (1979): 155–176.

19. See G. A. Cohen, "Reconsidering Historical Materialism," in *Marxist Theory,* ed. A. Callanicos (Oxford University Press, 1989), 148–74, esp. 149 note 3; Misa, "How Machines Make History," 319–322.

olution, and in causal historical analysis generally, machines make history when analysts adopt macro perspectives, whereas machines are *made by* historical processes whenever analysts adopt micro perspectives and strip machines of their ability to appear causative of social change.

Technology and Industrial Structure

A close analysis of a key structural change in the American mass-production steel industry will reveal the strengths (and weaknesses) of both approaches. For the analysis, I review a macro interpretation that stresses the transformative power of technology, then propose a micro interpretation that questions whether technology was causative of change. Portraying two opposing perspectives as equally plausible may seem churlish or worse; however, such an analysis is necessary if we are to assess the strengths and weaknesses of each perspective, thereby setting the stage for a synthesis of the two.

I make no special claim for the period under investigation—the second industrial revolution (1870–1920), comprising the familiar developments of science-based industry and the spread of modern corporate forms (among them the rise of vertical integration, the great merger movement at the turn of the century, and a transformation of the political economy).[20] Nor need we make any special allowance for the steel industry; along with the electrical, chemical, and manufacturing industries, its development was broadly typical of the period. Finally, although it superficially may appear so, I am not trying to demonstrate the historical "falsity" of macro interpretations while elevating the historical "truth" of micro interpretations. Although micro studies are the traditional stuff of the historian's craft, I take seriously McNeill's assertion that macro studies are possible and can be valid.[21]

20. Alfred D. Chandler, Jr., *The Visible Hand: The Managerial Revolution in American Business* (Belknap Press, 1977); Naomi R. Lamoreaux, *The Great Merger Movement in American Business, 1895–1904* (Cambridge University Press, 1985); Martin J. Sklar, *The Corporate Reconstruction of American Capitalism, 1890–1916: The Market, the Law, and Politics* (Cambridge University Press, 1988); James Livingston, "The Social Analysis of Economic History and Theory: Conjectures on Late Nineteenth-Century Development," *American Historical Review* 92 (February 1987): 69–95.

21. McNeill (*Mythistory,* p. 85) maintains: "Precision and truthfulness do not

The achievement of vertical integration at Carnegie Steel ensured that this firm, along with the others that made comparable moves to control raw materials, would dominate the adolescent steel industry. The majority of the firms that integrated vertically during the late 1890s wound up in the U.S. Steel combine (formed in 1901). The remaining integrated steel firms survived at the margins where U.S. Steel deliberately chose not to compete; non-integrated firms were relegated to specialty markets. Carnegie Steel's moves to control coke and especially iron ore triggered the most important industrial consolidation in the steel industry, leading directly to the formation of U.S. Steel.[22]

In discussing this period in American industry, Chandler allows that there can be several motives, besides cost-minimizing, for vertical integration or horizontal combination. He cites the desire to control competition by erecting barriers to entry, the advantages of picking up profitable property, the wish to maintain or create a tractable workforce, and the potential to manipulate securities. Yet, Chandler argues, combinations remained profitable and powerful in the long haul only if they rationalized the facilities acquired, integrated production with distribution, and created an extensive managerial hierarchy—and if the technology permitted high-volume mass production. In the United States, "technology and markets have been the basic determinants of the size of firms and of concentration in industry." Oliver Williamson, the theorist of markets and hierarchies, concurs: "Not all business executives, Andrew Carnegie included, accurately perceive their business opportunities and faultlessly respond. Over time, however, those integration moves that have better rationality properties (in transaction cost and scale economy terms) tend to have better survival properties."[23] Such a statement

necessarily increase as the scale becomes smaller. Large-scale truths and patternings can be just as precise as small-scale observations and truths."

22. This argument is developed in chapter 4 of *A Nation of Steel,* which also provides additional detail on the process of vertical integration at Carnegie Steel and other firms.

23. Alfred D. Chandler, Jr., "Markets, Hierarchies, and Hegemony: A Discussion," in *The Essential Alfred Chandler: Essays Toward a Historical Theory of Big Business,* ed. T. McCraw (Harvard Business School Press, 1988), 449, 460. Chandler's comment came during an interchange with the organizational sociologist Charles Perrow, who in countering both Chandler and Williamson had maintained the following (p. 433): "I will emphasize the control of

perfectly captures the logic and spirit of a macro perspective: abstracting from individual cases, imputing rationality on an actor's behalf, and being order-driven.

Chandler's discussion of vertical integration in the steel industry deduces from their actions that Carnegie Steel's executives were motivated by defensive considerations—i.e., considerations other than strict profit maximizing.[24] By the early 1890s (as Chandler relates it), Carnegie Steel's chairman, Henry Clay Frick, had consolidated and rationalized the several Carnegie manufacturing properties in and around Pittsburgh into an integrated whole.[25] He had also systema-

markets, the control of labor, and the social cost of undesirable sources of private profit. Any efficiencies that obtain through the reduction of the costs of transacting business (Williamson) or coordination (Chandler) are, I argue, minor consequences. They neither motivate entrepreneurs nor make a substantial contribution to profits. Profits come from control of markets and competition, control of labor, and the ability to externalize many other costs that are largely social in nature, that is, to force communities and workers to bear them and not have them reflected in the price of goods and services." This argument—between those who (like Chandler and Williamson) focus on allocative efficiency, and those who (like Perrow) focus on market power—is reformulated by Tim Knapp in "Hierarchies and Control" (*Sociological Quarterly* 30, no. 3 (1989): 425–440) and by Simon Ville in "The Expansion and Development of a Private Business: An Application of Vertical Integration Theory" (*Business History* (October 1991): 19–42).

24. The next three paragraphs closely follow Chandler's classic article "The Beginnings of 'Big Business' in American Industry" (*Business History Review* 33 (1959): 1–31; reprinted in *Essential Alfred Chandler,* ed. McCraw). A condensed account of vertical integration is given in Chandler, *Scale and Scope: The Dynamics of Industrial Capitalism* (Harvard University Press, 1990), pp. 128–131.

25. To my knowledge, there is no satisfactory account of Frick's rationalization of Carnegie's steel holdings during 1889–1892 (before the Homestead strike)—or, for that matter, during 1893–1899. Frick's biographer skips over this period in four vague paragraphs (George Harvey, *Henry Clay Frick—The Man* (Scribner, 1928), 93–94, 99). Joseph Wall's magnificent biography of Frick's boss devotes little attention to these changes (*Andrew Carnegie* (Oxford University Press, 1970; University of Pittsburgh Press, 1989), 497, 506). Early biographies of Carnegie say little about Frick; see, e.g., Burton Hendrick, *The Life of Andrew Carnegie* (New York, 1932; Harper & Row, 1969). The best account is in James Howard Bridge's *The Inside History of the Carnegie Steel Company* (Aldine, 1903); however, Bridge takes a number of pages (167–183, 295–297) to say surprisingly little about *what* Frick did. All commentators agree that Frick's most advantageous move for the Carnegie company

tized and departmentalized its purchasing, engineering, and marketing activities. Expanding the sales department became necessary, since a shift in the firm's focus from rails to structures had enlarged the number of its customers. Moreover, in 1896 the Carnegie company made a massive purchase of ore lands by joining with Henry Oliver to buy out the Rockefeller holdings in the Mesabi Range in north-central Minnesota. The depression of the 1890s had transformed the region. By 1896, the ore fields were dominated by three great interests (the Oliver Mining Company, the Minnesota Mining Company, and Rockefeller's Consolidated Iron Mines); a fourth, James J. Hill's Great Northern Railroad, was just entering the field. Frick's purchases ensured that the Carnegie company would have cheap ore and provided it with a fleet of ore ships. Next, Frick and Carnegie bought and rebuilt railroads from Lake Erie to Pittsburgh to carry the new supplies to the mills.

Yet, continues Chandler, the Carnegie company's managers did little to coordinate the various mining, shipping, and manufacturing units. These activities did not become departments controlled from one central office; they remained separate companies under independent managements, whose principal contact with one another was through negotiated contracts. A similar relation existed between the Frick Coke Company and Carnegie Steel from the early 1880s onward; indeed, it continued even after Carnegie named Frick as the steel firm's chairman in 1889. If the Carnegie company's strategy had been to provide a more effective flow of materials as well as to ensure that it would not be caught without a supply of ore and the means to transport it, Chandler maintains, Frick and Carnegie would have created some sort of central coordinating office.

Other steelmakers responded quickly to the Carnegie purchases. In 1898, Chicago's Illinois Steel Company (with capital supplied by J. P. Morgan) joined the Lorain Steel Company (with plants on Lake Erie and in Johnstown, Pennsylvania) to purchase the Minnesota Mining Company, a fleet of ore boats, and railroads in the Mesabi and Chicago areas. Again, little attempt was made to coordinate mining and shipping with manufacturing and marketing. Also in

during 1889–1892 was his purchase of the Allegheny Bessemer Steel Company (later the Duquesne Steel Company)—a financial transaction, rather than managerial rationalization. In "The Beginnings of 'Big Business'" Chandler's sources are Hendrick, Harvey, and Bridge.

1898, a number of iron and steel firms in Ohio and Pennsylvania merged to form the Republic and National steel companies; shortly thereafter, a similar combination in the Sault Saint Marie area became the Consolidated Lake Superior Company. These three new companies began at once to set up marketing organizations and to obtain control, by lease and purchase, of raw materials and transportation facilities. In these same years, the large established steel companies—Lackawanna, Cambria, Jones & Laughlin—obtained control of more supplies of ore, coke, and limestone and simultaneously reorganized their manufacturing and marketing organizations. Like Carnegie and Federal, they at first made little effort to bring their mining and coke operations under the direct control of the central office.

Chandler has abstracted a general pattern from the case studies. Moreover, he has provided a variant of the macro method of imputing rationality. The actors' motives are inferred from their actions rather than from their testimony; the conclusion is that, since their actions were not to centralize control of the mining operations, their motives must have been defensive. Finally, the entire program of charting the emergence of managerial hierarchies is thoroughly order-driven. Needless to say, Chandler's actors, here and elsewhere, are rational, purposeful agents of change.

A micro-level interpretation of this period searches for the timing and the motives (as best as they can be ascertained) of the steps later perceived as vertical integration. If vertical integration is defined as purchasing (or controlling) additional units upstream or downstream from the core manufacturing stage, the process began not in 1889, when Frick came on board, but in 1871, when Andrew Carnegie formed a subsidiary company to build a blast furnace to supply pig iron for his Union Iron Mills. A more significant step toward vertical integration occurred with the 1881 reorganization that founded Carnegie Brothers & Company; at this time the Edgar Thomson works (opened in 1875) lost its separate identity and joined the Union Iron Mills, the Lucy Furnaces, the Scotia ore mines, and several Carnegie-owned coal mines and coke ovens in the Connellsville district. The 1881 consolidation appears to have been motivated not by considerations of efficiency but by fallout from "the internal discord in which all the Carnegie enterprises were born and brought up."[26] Profitable investments in the securities of railroad, bridge, and iron-rolling

26. Bridge, *Inside History*, 117–35, quote 117; Wall, *Andrew Carnegie*, 471–72.

ventures, as well as the commissions for selling these securities in New York and London, gave Andrew Carnegie the financial wherewithal to make major purchases. But it was the bitter infighting among his partners and among certain of their competitors that gave Carnegie the opportunity to buy up shares in all these partnerships and consolidate them under his aegis. James H. Bridge observes that "it was other considerations than increased efficiency and economy that prompted the first imperfect combination of the Carnegie properties."[27] Moreover, vertical *dis*-integration occurred within 2 months of the selling of the Lucy furnaces. Joseph Wall writes that "at this point neither Carnegie nor his associates saw the advantages of having within one company the entire process, from the mining of raw iron ore and coal through the manufacture of pig iron to the converting of pig iron into steel."[28]

Five years later in 1886, when the annual net profit reached 60 percent of the total original capitalization of the firm, Andrew Carnegie effected another major reorganization. He organized a second firm as Carnegie, Phipps & Company (which held the Homestead mill, purchased in October 1883, and the recently repurchased Lucy Furnaces) while retaining Carnegie Brothers & Company as a separate entity.[29] Again, as best as can be ascertained, the motivations for acquiring the Homestead mill—a state-of-the-art facility built in 1879 by the Pittsburgh Bessemer Steel Company to produce Bessemer steel rails—were prescient opportunism and the desire to buy out "a dangerous rival" rather than any strategic vision of vertical integration. Again, the occasion of the purchase was infighting among the partners of the Homestead mill compounded by labor unrest that had culminated in a ten-week strike in the spring of 1882. Once in the Carnegie company's control, Homestead was converted to the production of structural steel shapes, an important diversification for the industry as a whole; the extent to which this new market for structural shapes *required* compensating organization change remains, in the absence of a detailed account of this period, a matter of deduction and conjecture. An odd consolidation it was. The anomaly

27. Bridge, *Inside History,* 117–35, quote 135; Wall, *Andrew Carnegie,* 471–72.

28. Wall, *Andrew Carnegie,* quote 472; Bridge, *Inside History,* 135, concurs: "The advantages of industrial consolidation had not, at this date, received any general recognition."

29. Wall, *Andrew Carnegie,* 472.

of creating two parallel firms in 1886 reflected Carnegie's fear of creating a single firm with heavy capitalization as well as his desire to maintain the purely financial advantages of having two separate firms.[30] The buildup of the Carnegie enterprises "was a natural and unconscious development growing out of trade conditions," writes Bridge. "There was at no time a well-defined plan or policy of expansion."[31]

Considerations of mortality were also a significant motivation for organizational change. During the fall of 1887, Andrew Carnegie's younger brother Tom died of pneumonia, while Andrew himself lay sick with typhoid fever. Tom Carnegie was not only the firms' next-to-largest shareholder but also a capable manager and a trusted adviser—roles that were especially important in that Andrew had directed his Pittsburgh operations from his home in New York for 20 years. The firms barely recovered from the loss of Tom Carnegie and they would not have survived the loss of his brother. As privately held, limited partnerships, they were particularly susceptible to loss of partners. The 1887 crisis shocked the firms into creating the "Iron Clad" agreement, which provided that the company could purchase for itself at book value any partner's share in the case of the partner's death or (as later amended) the case of a three-fourths vote that any partner's continued presence was detrimental to the partnership.[32]

No special strategic vision seemed to motivate Carnegie's offer to Henry Clay Frick, on January 31, 1887, of a 2-percent share of Carnegie Brothers' capital for the book value of $184,000. Like other favored junior partners before him, Frick paid for his share over

30. Wall, *Andrew Carnegie,* 475–76, 486–89, 535; Harvey, *Henry Clay Frick,* 101–2; Bridge, *Inside History,* 150–69. Working capital was obtained by discounting notes given by one company to another, as buyer and seller, thus providing the "two name paper" required by banks; in fact, there emerged a three-cornered arrangement among the twin Carnegie companies and the Frick coke company; in 1890 this technique was used to issue $2,635,000 in these notes.

31. Bridge, *Inside History,* quote 168. Bridge ascribes to Frick, from January 1889 chairman of Carnegie Brothers & Company, the rationalization of this firm; in January 1889 Frick was also elected a director of Carnegie, Phipps & Company but (according to Harvey, 101) "took no part in administration."

32. Voting was by number of shares each person held; since Andrew Carnegie held over half the total shares, he could not be ousted by any combination of his partners.

time through its declared dividends. Like others, Frick signed the Iron Clad; unlike others, though, Frick would be forcibly ejected from the Carnegie empire through the Iron Clad.[33] The partnership did culminate an alliance between the Pittsburgh district's most prominent steelmaker and its most aggressive coke producer, however. Through the 1880s, the Carnegie firms provided handsome financing for Frick to expand his coking operation in the Connellsville region, displacing the Pittsburgh banker Judge Thomas Mellon as Frick's chief source of capital and emerging as the majority owner of Frick's expansive coke empire. This partnership completed one phase of vertical integration (control of coal and coke). Carnegie, for his part, lauded Frick's talents as a hard-boiled, cost-cutting manager, and in January 1889 elevated him to chairman of Carnegie Brothers & Company, increasing his share from 2 to 11 percent; but Carnegie still did not appreciate the strategic importance of vertical integration for nearly another decade.[34] Indeed, for the second phase of vertical integration, "Carnegie's junior partners would have to prod him into risking capital for the acquisition of iron ore."[35]

The acquisition of the nation's richest iron-ore lands was effected by the Carnegie Steel Company, Limited, formed July 1, 1892. The new firm's capitalization of $25 million—2.5 times the twin firms' combined capitalization—reflected remarkable growth since the 1886 reorganization. At last the Carnegie enterprises were collected into one organization; and over the organization presided H. C. Frick. Ironically, before the privately held company was a week old, the bitter Homestead strike would put it in the public eye and into the annals of labor history.[36]

With the Carnegie firm's acquisition of iron-ore lands in the mid 1890s, one can for the first time identify actors motivated by the strategic advantages of controlling an essential raw material. Nonetheless, this outcome did not reflect the single-minded motivation of

33. Wall, *Andrew Carnegie,* 489–93.

34. Harvey, *Henry Clay Frick,* 76–92; Wall, *Andrew Carnegie,* 478–86, 497.

35. Wall, *Andrew Carnegie,* quote 587.

36. Many accounts of the 1892 Homestead strike exist; for a new labor history perspective see Linda Schneider, "The Citizen Striker: Workers' Ideology in the Homestead Strike of 1892," *Labor History* 23 (1982): 47–66; Paul Krause, *The Battle for Homestead, 1880–1892: Politics, Culture, and Steel* (University of Pittsburgh Press, 1992).

any of the Carnegie partners—not even Frick, who had done so much to build up an empire in coke and deliver it to the Carnegie realm. Carnegie had a saying that Fortune knocked once at every man's door; in this case, Fortune—the richest deposits of iron ore in the nation—knocked time and again before Carnegie permitted entry. "It is a story," Bridge observed, "that shatters all preconceptions of the genius necessary to achieve millionaireship."[37]

By the early 1890s, the insatiable maw of the nation's blast furnaces had drawn down the iron-ore deposits of the Marquette, Menominee, and Gogebic ranges in Michigan's upper peninsula, and ore prospectors turned their attention to the newly opened Vermilion and Mesabi ranges. As early as 1865 Minnesota's geologist had reported finding "immense bodies of the ores of iron" on the eastern Mesabi, but not until the Merritt family explored this region beginning in 1888 was this observation systematically followed up. In 1890, based on their mappings, the Merritt brothers obtained 155 leases on mineral rights of up to 160 acres each; by the spring of 1891, when Lon Merritt set out to rally support in Pittsburgh, the brothers had joined with Duluth investors to establish two companies with a total capitalization of $4 million. Frick's reception of Merritt was cool in the extreme, but before he was ejected from Frick's office Merritt did persuade Frick to dispatch his leading expert to look over the range. The expert knew the hard ores of the established ranges; seeing instead the soft granulated substance characteristic of the Mesabi range, he "swore it was not iron ore" and declined to have a sample assayed in Duluth. Had he done so, he would have discovered pure Bessemer-grade ore, 50 to 65 percent iron with but a trace of harmful phosphorus.[38]

Fortune would knock again on Carnegie's door in the person of Henry Oliver. Oliver, a childhood friend of Carnegie who had entered the steel industry in the early 1880s after nearly two decades promoting Pittsburgh railroads and iron manufactures, first visited the Mesabi range during the summer of 1892 after attending the Republican National Convention in Minneapolis. He headed north, saw the efforts of the 51 mining companies then sending out iron

37. Bridge, *Inside History,* quote 257, 268.

38. David A. Walker, *Iron Frontier: The Discovery and Early Development of Minnesota's Three Ranges* (Minnesota Historical Society Press, 1979), 73–100, quotes 74, 100; Wall, *Andrew Carnegie,* 587–94.

ore from the Mesabi to the harbor at Duluth, convinced the Merritt brothers with a $5000 check that the Carnegie steel companies would weigh in with big investment, then hastened back to Pittsburgh to make good his promise. Frick now favored the deal Oliver proposed: a one-half interest in Oliver's mining company (which he had organized with the Merritts) for a loan of $500,000, secured by a mortgage on the ore properties. But Carnegie, in a letter from Scotland dated August 29, 1892, demurred: "Oliver's ore bargain is just like him—nothing in it. If there is any department of business which offers no inducement, it is ore. It never has been very profitable, and the Massaba [i.e. Mesabi] is not the last great deposit that Lake Superior is to reveal." After the settlement of the Homestead strike, Frick successfully concluded the deal with Oliver, and the Carnegie firm took up its half share in the Oliver firm. Twenty months later, in a letter to his board of managers dated April 18, 1894, Carnegie still denied the fortune that Oliver persistently dangled before him: "The Oliver bargain I do not regard as very valuable. You will find that this ore venture, like all our other ventures in ore, will result in more trouble and less profit than almost any branch of our business. If any of our brilliant and talented young partners have more time, or attention, than is required for their present duties, they will find sources of much greater profit right at home. I hope you will make a note of this prophecy."[39] Many, indeed, delighted in Carnegie's stunningly wrong prophecy.

By this time, however, forces as powerful as Carnegie's firm and twice as determined had moved onto the Mesabi range. Indeed, the transformation of the Mesabi in these years from an entrepreneurial venture run by families to a rationalized empire controlled by corporations nicely encapsulates the merger movement that transformed American industry between 1895 and 1904. For the Mesabi entrepreneurs, the 1893 panic and the subsequent depression had been a disaster. Orders for iron ore from steel mills vanished just as the Merritts' credit collapsed. Into the breech stepped John D. Rockefeller, who in 1894 took over the major share in the newly formed Lake Superior Consolidated Iron Mines Company, which owned the

39. Bridge, *Inside History,* 257–59, quotes 259; Wall, *Andrew Carnegie,* 594–97; Walker, *Iron Frontier,* 101–18; Hendrick, *Life of Andrew Carnegie,* II: 12–14; Terry S. Reynolds, s.v. "Henry William Oliver," in *Iron and Steel in the Nineteenth Century,* ed. P. F. Paskoff (Facts on File, 1989), 259–62.

six mines the Merritts had opened and held independently from Oliver. Rockefeller's activities induced Carnegie to increase his share in Oliver's mining company from one-half to five-sixths. Though contemporaneous speculation had it otherwise, Rockefeller saw the Mesabi mostly as a source of freight for his transportation network, and by late 1896 he was willing to strike a mammoth deal with Carnegie. For 50 years, the Oliver Mining Company and Carnegie Steel were to lease all of the ore property of the Consolidated company, paying a royalty of 25¢ per ton extracted and guaranteeing a minimum purchase of 600,000 tons of ore each year. Rockefeller was to gain the traffic on his railroad lines and lake steamships of this 600,000 tons plus an additional 600,000 tons from the Oliver properties, and each side agreed not to enter the other's domain (thus Rockefeller's Mesabi ore holdings and Carnegie's steelmaking empire were protected from each other). The agreement in effect slashed the going ore royalty from 65¢ per ton to 25¢ per ton; its consequences swept across the entire ore-producing region: "This alliance with the Rockefellers . . . produced a panic among the other mine owners; and stockholders in Boston, Chicago, Cleveland, and the Northwest hastened to get rid of their ore properties at almost any price. The demoralization extended to the ore markets; and Norrie [ore from the largest mine on the Gogebic Range], which sold at $6 a ton in 1891, dropped to $2.65 on the docks at Cleveland."[40]

With ore properties in a doubly depressed condition, Henry Oliver labored mightily to deliver Fortune to Carnegie's door. Carnegie had directed Oliver to make preparations to deliver a full 2,000,000 tons of iron ore during 1897, and Oliver soon translated this into a mandate for getting options on ore properties—whose owners, numbering in the hundreds, were all too willing to sell. The 1896 agreement had barred the Carnegie concern from Rockefeller's domain of the Mesabi, so Oliver turned to the other principal ore regions; on July 27, 1897, he reported to Frick that they could acquire by leasing or purchase three important mining areas—including the Norrie mines—in the Gogebic and Vermilion ranges. These new mines would add 1,600,000 tons of ore to the Mesabi's 1,200,000 tons

40. Bridge, *Inside History,* 258–60, quote 261; Wall, *Andrew Carnegie,* 598–601; Walker, *Iron Frontier,* 127–210; Reynolds, "Henry William Oliver," 262–63.

shipped each year. Oliver saw the strategic advantages that would accrue to the firm, and he articulated them to Frick:

> I propose at a risk of using our credit to the extent of $500,000, or possibly one million dollars, to effect a saving, in which our competitors will not share, of four to six million dollars per annum. . . .
>
> I desire to impress upon you the fact, that if it had not been for our Rockefeller—Mesaba deal of last year, with the consequent demoralization in the trade caused by the publication thereof, it would not have been possible for us to now secure the other Range properties I propose to acquire, either by lease or for any reasonable price. We simply knocked the price of ore from $4.00 down to say $2.50 per ton. Now let us take advantage of our action before a season of good times gives the ore producers strength and opportunity to get together by combination.

Carnegie—reluctantly, it seems—agreed to part of Oliver's proposal, but he flatly rejected the Norrie mines, the largest of the three properties. "Policy of firm not to invest Capital in mining, but to lease mines upon favorable terms," cabled Carnegie to Charles Schwab, the Carnegie firm's new president, who favored purchasing the entire set of properties. In part, Carnegie was worried about potential trouble with the Norrie's minority shareholders; but more fundamentally, Carnegie was fixated on short-term financial calculations suggesting that the rate of return on capital invested in the ownership of ore properties would not be as high as the rate of return in manufacturing. Frick sided with Carnegie against purchasing the Norrie mines. Not even Rockefeller was in the market for additional ore properties. As summer turned to autumn, Oliver, whose carefully arranged options were to expire on September 30 became increasingly desperate. With five days to go, he cabled a last-ditch appeal to Carnegie, who relented his steadfast opposition and cabled his go-ahead to the board of managers: "Always approve unanimous action of board after full expression views. Sure leasing true policy but if board decides this exception all right." The day Carnegie's message was received, the board met in emergency session and unanimously approved the purchase of the Norrie mines. Frick, who had changed his mind in the nick of time, cabled Carnegie their positive decision and concluded: "This removes the Carnegie Steel Company, Limited from the Bessemer Ore Market. Business broadening; outlook good." Only after this juncture was Carnegie convinced of the benefits of owning or controlling iron ore mines, and then Oliver went to work. By the end of 1899, he and Carnegie Steel controlled through lease or ownership 34 working mines across the ore-producing region and 16 exploratory sites on all the ranges except the Mesabi. When the Carnegie-Oliver properties were folded into the United States Steel Corporation, in 1901, they represented two-thirds of the known northwestern supply of Bessemer ores—approximately 500 million tons, valued at upwards of $500 million.[41]

It is tempting to see an ironic parallel between how Carnegie gained an empire in steel and how some say England gained an empire in the New World: "in a fit of absence of mind." At the least, the rational, order-seeking pattern of vertical integration owes more to historians' analyses than to contemporary actors' perceptions. Not the Carnegie Steel executives, least of all Andrew Carnegie, but Henry Oliver first imagined the long-term strategic advantages to the steel firm of owning and controlling its ore supply. The challenge for a micro interpretation is to explain the paradoxical result by which a nonrational process (acquisitions made for motives unrelated to the process of vertical integration) eventually yielded a rational result. The response is a methodological observation: The emergence of any particular pattern, especially across a significant duration of time, does not imply that any single individual's motive must have been to seek that pattern. Technological-determinist accounts often conflate outcomes (long-term patterns that can be entirely independent of actors' conscious agendas) with motivations (short-term conscious agendas of actors). To assert that the emergence of a particular pattern was *caused* by someone's consciously planning the emergence of that pattern can be, in the absence of evidence, to commit the logical error of *post hoc ergo propter hoc.*

Retrieving Sociotechnical Change from Technological Determinism

It is plausible, given the foregoing micro-level analysis, to see technological change as a necessary but not sufficient cause for change in the industrial structure of steelmaking. Competing in the mass production of steel simply required Bessemer technology after the 1870s, plant-wide integration after the 1880s, and vertical integration after the 1890s. For structural changes, however, other motives were

41. Wall, *Andrew Carnegie,* 602–7, quote 603–4; Walker, *Iron Frontier,* 211–16; Bridge, *Inside History,* 268. Oliver's 27 July 1897 report to Frick is printed in Bridge, *Inside History,* 261–66.

necessary.[42] Vertical integration proved economically rational (lower costs for iron ore plus barriers to entry) for Carnegie Steel, but it was a two-decade-long process whose rationality appeared most forcefully to later historians and not to the actors themselves. "Fortunately," Andrew Carnegie conceded about his firm's protracted effort to control iron ore, "I woke up in time."[43] As we have seen, technological-determinist accounts remain plausible only when the analysis stays at the macro level; micro-level analysis dissolves the technical logic, necessity, and order-driven patterns into a swirl of competing personalities and contingent factors.

In view of the above comments, it is essential to emphasize that the foregoing analysis is not a partisan attack on macro studies. On the one hand, macro studies may be faulted for leveling historical processes that are full of conflicting values and actors, and for hampering the integration of macro and micro levels by invoking deterministic forces that micro studies literally cannot locate.[44] "The central notion for all varieties of macro-history," affirms William McNeill, "is that of a social process (or processes) acting largely in independence of human awareness and so, by definition, not to be found recorded and awaiting discovery in some primary archive." On the other hand, micro studies, in the attempt to demonstrate the socially constructed nature of technology, often omit comment on the intriguing question of whether technology has any influence on anything. This result (however unintended) seems especially undesirable in an age of pervasive sociotechnical problems. McNeill writes: "Interaction between large views, bold hypotheses, fallible intuitions and exact, detailed

42. Compare Chandler in McCraw, *Essential Alfred Chandler,* 463: "long-term profit was the underlying motive for the growth of firms. . . . More specific motives were involved in achieving that goal."

43. Carnegie testimony to the Stanley Committee in 1912, quoted in Hendrick, *Life of Andrew Carnegie,* II: 14.

44. The Wittfogel thesis (discussed in this volume by Peter Perdue)—which links the need for centralized control of water supplies in Oriental societies to the inevitable development of autocratic political regimes—is one such macro analyst's abstraction; as such, it is unlikely that any micro analyst's evidence will confirm it. Alfred Chandler's "minimum efficient scale" (defined as the scale of operation necessary to reach the lowest cost per unit) is another such macro concept; see Chandler, *Scale and Scope: The Dynamics of Industrial Capitalism* (Harvard University Press, 1990), 23–24.

scholarship is what we need."[45] Of course. But if macro-level outcomes and processes are by definition independent of micro-level motivations and actions, where are historians seeking common ground to look?

My prescription for synthesizing the social-construction and society-shaping theses flows from the above analysis. Historians must recognize that what we choose to study, as well as on which side of the macro/micro divide it lies, strongly influences whether technology emerges as socially constructed or as society-shaping. To heal this methodological bifurcation, and hence formulate a new and more insightful analysis of technology and social change, historians should direct attention to what can be called the "meso" level—the region conceptually intermediate between the macro and the micro. For historians of technology and business, this means analyzing the institutions intermediate between the firm and the market or between the individual and the state. A short list of these might include manufacturers organizations (including cartels and interfirm networks), standard-setting bodies (including the engineering profession and public agencies), export-import firms specializing in technology transfer, consulting engineering firms, and investment banking houses.[46] Since these institutions mediate between key actors in soci-

45. McNeill, *Mythistory,* 50–51, 128.

46. Some historians, sociologists, and economists have already taken up this task, though not always with the awareness that they are studying the meso level. On international technology transfer see Geoffrey Tweedale, *Sheffield Steel and America: A Century of Commercial and Technological Interdependence, 1830–1930* (Cambridge University Press, 1987) and Kristine Bruland, *English Technology and European Industrialization* (Cambridge University Press, 1989); on consulting engineering firms and investment banking houses see Thomas P. Hughes, *Networks of Power* (Johns Hopkins University Press, 1983); on standard-setting bodies see Håkon With Andersen and John Peter Collett, *Anchor and Balance: Det Norske Veritas, 1864–1989* (Oslo: J. W. Cappelens Forlag, 1989) and Susan J. Douglas, *Inventing American Broadcasting, 1899–1922* (Johns Hopkins University Press, 1987).

For the recent literature on cartels, another intermediary institution, see Thomas S. Ulen, "Railroad Cartels before 1887: The Effectiveness of Private Enforcement of Collusion," *Research in Economic History* 8 (1983): 125–144; C. Knick Harley, "Oligopoly Agreement and the Timing of American Railroad Construction," *Journal of Economic History* 42 (December 1982): 797–823; Daniel Barbezat, "A Price for Every Product, Every Place: The International Steel Export Cartel, 1933–1939," *Business History* 33 (October 1991): 68–86; Richard A. Lauderbaugh, "Business, Labor, and Foreign Policy: U.S.

ety, whether they orchestrate or respond to sociotechnical change, such an analysis would naturally lead us to the historical public debates concerning the costs and benefits of sociotechnical change. It would thus provide a framework in which to investigate who paid the costs for efficiency and growth as well as who engineered the efficiency in the first place.

Finally, what can we say about technology and social change—besides the observation that historians' conclusions depend critically on the perspectives they adopt? On the one hand is the proposition that technological change has deterministic effects on social processes, which, I think, we should continue to reject on the evidence of micro-level studies. On the other hand is the charting of macro-level social processes that exist independent of contemporaneous individuals' consciousness. Contemplating this uncertain terrain, most historians will hesitate before giving up their ability to say something about individuals' motivations. One suggestion of the above analysis is that historians should move between these two levels, but that they must take care not to repeat the methodological error of confusing the long-term emergence of patterns at the macro level with the short-term motivations of actors at the micro level.

The perspective advocated here—a focus on the meso-level institutions and organizations that mediate between the individual and the cosmos—offers a framework for integrating the social shaping of technology and the technological shaping of society. Such a perspec-

Steel, the International Steel Cartel, and Recognition of the Steel Workers Organizing Committee," *Politics and Society* 6 (1976): 433–457; Steven B. Webb, "Tariffs, Cartels, Technology and Growth in the German Steel Industry, 1879 to 1914," *Journal of Economic History* 40, no. 2 (1980): 309–329; Lon L. Peters, "Are Cartels Unstable? The German Steel Works Association before World War I," *Research in Economic History*, supplement 3 (1984): 61–85.

Social scientists writing on these themes include Bengt-Åke Lundvall ("Innovation as an Interactive Process: From User-Producer Interaction to the National System of Innovation," in *Technical Change and Economic Theory*, ed. G. Dosi et al. (Pinter, 1988)), Hans Radder ("Experiment, Technology and the Intrinsic Connection between Knowledge and Power," *Social Studies of Science* 16 (1986): 663–683), Knut Sørensen and Nora Levold ("Tacit Networks, Heterogeneous Engineers and Embodied Knowledge," *Science, Technology and Human Values* 17 (1992): 13–35), and the contributors to *The Social Construction of Technological Systems*, ed. W. E. Bijker et al. (MIT Press, 1987) (see especially the essays in part III).

tive recognizes that a technology is far more than a piece of hardware. Properly understood, "technology" is a shorthand term for the elaborate sociotechnical networks that span society. To invoke "technology," on the macro level of analysis, is to compact into one tidy term a whole host of actors, machines, institutions, and social relations. To expand "technology," on the micro level of analysis, is to regain the complexity and messiness of the compacted whole. Insofar as people are necessary parts of the networks, to say that "technology" causes social change is really to say that people—through the sociotechnical networks they create and sustain—cause social change. Explaining, understanding, and managing these networks is the task before us.

Acknowledgments

I am indebted to several participants in the workshop "Machines and History: The Question of Technological Determinism" (Dibner Institute for the History of Science and Technology, MIT, December 2–3, 1989), including William McNeill, Peter Perdue, Alfred Chandler, and Carl Kaysen, for challenging or confirming several points of my analysis. I also received valuable feedback while presenting preliminary versions at the Institute for History of Technology, Technische Universität—Berlin; the Center for Technology and Policy Studies, Apeldoorn, Holland; and the Centre for Technology and Society, University of Trondheim. In addition Paul Barrett, Bernard Carlson, Thomas Hughes, Annemarie Mol, Arie Rip, Alex Roland, Johan Schot, Francis Sejersted, and Edmund Todd have each clarified my thinking in one way or another.

Determinism and Indeterminacy in the History of Technology

Philip Scranton

Philip Scranton makes a strong case against deterministic master narratives and in favor of more variegated contextual accounts that recognize the complexity and indeterminacy of historical processes. He maintains that American political-economic hegemony and American history are currently in disarray. The former, governed by visions of infinite progress through technology, reached its apogee in the post-World War II period only to erode in the face of both increasing foreign competition and national traumas such as the Vietnam War. The latter, doubtless influenced by developments in the first arena, has resulted in the ghettoization of history as scholars with "incommensurable approaches carve out fiefdoms in the imaginary terrain of subdisciplines." As a way of reconfiguring the discipline, Scranton proposes that historians set "totalizing determinism aside" to probe what he refers to as "local determinations" in which technology is construed as part of larger sociocultural processes. While urging historians to be attentive to the theoretical implications of their work, he nonetheless cautions them to "steer clear of fresh reductionisms" by taking into account the contingencies, diversities, disjunctions, multiple oppositions, and contrasting norms that often get pushed aside in deterministic narratives. Scranton's avowedly postmodernist argument contrasts sharply with the positions of Robert Heilbroner and Thomas Hughes. In effect, he presents historians of technology and science with a challenging agenda for the twenty-first century.

In a society committed rhetorically and mythically to the significance of the individual, voluntarism, and agency, determinism is a discomfiting notion. In their strong forms, be they Calvinist or crude Marxist, determinisms calls forth images of universal structures and dynamics that deny or sharply delimit the capacities of individuals and institutions to alter history's trajectory on the personal, the social, or the global plane. People or nations become featured or bit players in transcendent dramas, thereby gaining stable meanings and roles, rules for practice, and confidence in outcomes. If Puritan anxiety, flashes of antinomianism, or waves of proletarian optimism materialize along the way, incurring human costs, they are easily enough slotted into the master narrative.

The externality and otherness of such a governing framework ill suited self-made men dedicated to conquering nature and natives in order to design and erect the political economy of the United States. Here agency ruled, theocracy was rejected (and scaled down to a market democracy of sects), and universals were recalibrated spatially, recoded nationally, and adjusted periodically across the span from manifest destiny to the era of global policing. If liberty was the core organizing value, its meaning was far from fixed and its endurance far from certain. Fragile and beset by contingencies, it had to be defended actively, extended cautiously and aggressively by turns, and revitalized repeatedly. National vitality flowed from appreciating this vulnerability and translating that understanding into action. It was not guaranteed by a cosmic or historical logic.

Yet underneath this dismissal of strong determinisms lay acceptance of a softer assurance. American greatness (not human perfectability) would arise from the mobilization of boundless resources and the application of technical ingenuity, creating a continent-wide variant of cumulative progress through practical rationality. Despite their differences over the ideal path and its political implications, Jefferson and Hamilton shared such a sensibility—one that was welded to incommensurate notions of liberty and to common worries for liberty's future. Divergent regional practices led Jefferson's commercial husbandry to root in the Middle West, whereas Hamilton's manufacturing vision was realized in the North, with the South an isolate, extending westward yet conserving the "land-killer" strategies that propelled such movement.[1]

1. Gavin Wright, *Old South, New South* (Basic Books, 1986), pp. 30–31.

The Civil War confirmed both the South's error and the larger faith that technical advances at farm and factory undergirded the defense and elaboration of American liberty. That this line of determination was neither universalized nor naturalized is evident from the long devotion to tariff protection for industry and import-menaced agriculturalists (e.g. wool producers), stretching from Morrill to Smoot-Hawley. The reigning "general theory" in political economy, free trade, sprouted often in national debates but never in policy.[2] Incremental and breakthrough innovations, together with an array of transatlantic borrowings, built prodigious American technical capacity and confidence across the half-century following the nation's centennial. From a nationally bounded and politically informed expectation, technical progress arguably emerged as the symbolic engine propelling American eminence, then preeminence.

If half the world's cars were Fords in 1920,[3] if American industrial might had turned the tide of the Great War, if its second-industrial-revolution giants were surging toward global leadership, it took little effort to shun an earlier sense of contingency. Technological momentum seemed inexorable, firing the imaginations of Thomas Hughes's system builders, rendering the Great Depression an unfortunate misstep on the road to the American Century, and solidifying what John Staudenmaier has termed a "confidence in the gradual triumph of Western rationality and the incremental advance of scientific and technical progress."[4] If public intellectuals like Mumford and Chase made cautionary noises about complexity, risks, and downsides to these gains, government, big business, and industrial unions paid them no heed, crafting policies and expectations based on extrapolations of midcentury trends toward American dominance in critical civilian and military technologies and mastery of the associated organizational strategies. Progress became product, the United States its exemplar, and technological prowess its foundation.

2. Frank Taussig, *Free Trade, the Tariff, and the Reciprocity* (Macmillan, 1920), chapters 1 and 8.

3. Alfred Chandler and Richard Tedlow, *The Coming of Managerial Capitalism* (Irwin, 1985), p. 239.

4. Thomas Hughes, *American Genesis* (Penguin, 1989); John Staudenmaier, "Recent Trends in the History of Technology," *American Historical Review* 95 (1990): 725.

My conjecture is that technological determinism came closer to being an article of faith in the United States between 1940 and 1960 than ever before or since. Technological innovation powered the national economy, brought the consumer society to full blossom, and reinforced the American image at home and internationally. Our rivalry with the Soviet Union meant first bigger and better bombs, then faster and finer means of delivery (planes, missiles, nuclear submarines). We raced to the moon, and we ringed the planet with satellites for surveillance, targeting, and eventually for-profit broadcasting and communication. Yet if Richard Nixon was right, America's triumph was equally guaranteed by the household technologies he championed in the 1959 "kitchen debate."[5] That the extrapolations faded into stagflation, that the military hardware failed to ensure victory in Viet Nam, and that other nations soon after matched or bettered our technical performance in many commercial product lines brought quite a series of shocks. When (as Michael Smith has explained[6]) these shocks were aggravated by reactor failures, disarray in the space program, and wide evidence of environmental and health damage as unacknowledged or unanticipated consequences of progress, criticism of the responsible institutions sharpened and disenchantment with the cultural hegemony of technically driven and defined development set in.[7] The reawakened recognition that other elements in the socio-economic matrix (politics, market failure, organizational structure) powerfully conditioned the pattern of outcomes stimulated a host of seminal studies in the history of technology and related fields.[8]

5. Paul Boyer, *By the Bomb's Early Light* (Pantheon, 1985); Merritt Roe Smith, ed., *Military Enterprise and Technological Change* (MIT Press, 1985); Elaine May, *Homeward Bound* (Basic Books, 1988), chapter 1; Clark Clifford, "Serving the President," *New Yorker* (April 1, 1991): 73.

6. See Smith's essay "Recourse to Empire" in this volume.

7. On cultural hegemony see Jackson Lears, "The Concept of Cultural Hegemony: Problems and Possibilities," *American Historical Review* 90 (1985): 567–593, and Richard Butsch, ed., *For Fun and Profit* (Temple University Press, 1990), chapter 1. See also Charles Perrow, *Normal Accidents* (Basic Books, 1984), and Thomas Dunlap, *DDT: Scientists, Citizens, and Public Policy* (Princeton University Press, 1981) for unexpected consequences.

8. David Noble, *America By Design* (Knopf, 1977) and *Forces of Production* (Knopf, 1984); Langdon Winner, *Autonomous Technology* (MIT Press, 1977); Larry Hirschhorn, *Beyond Mechanization* (MIT Press, 1984); Max Holland, *When the Machine Stopped* (Harvard Business School Press, 1989).

In this setting, those who might wish to cleave to the classic determinism—the "Whig reading of Western technological evolution as inevitable and autonomous"[9]—could resort simply to decoupling it from the U.S. case, making the self-realization of technical efficiency a spatially movable feast.[10] Others, taking account of accumulating negatives, could press toward adapting public-policy methods to a historical cost-benefit analysis of technological change, expecting a positive bottom line. For historians of technology, in my judgment, neither maneuver is especially appealing. The first leads into a metaphysics of technology as a force manifesting itself in and through history, leaving nations, firms, and individuals dependent on detecting whether the force is with them. Restoration of a master narrative[11] vitiates reckoning with agency, indeterminacy, the overabundance of information,[12] and the battery of "external" activities that bear on the course of technical change. In my view, this tactic depends on the dubious practice of revising an older linearity rather than displacing it. The second option demands speculative quantification on a heroic order—the kind of data-churning involved in recent dollar valuations of pollution and cancer risk, which in turn have yielded public-policy initiatives that reduce social choices to a set of imaginary

9. Staudenmaier, "Recent Trends," p. 725.

10. As with Perdue's "determined determinists" in the case of agricultural development. (See Perdue's essay in this volume.)

11. For a brief and accessible discussion of the master narrative dilemma, see "Interview with Joan Scott," *Radical History Review* 45 (1989): 49–55. Thomas Misa's strategy of unpacking master (which he calls "macro") narratives by means of contrasting micro-level narratives (see his essay in this volume) leads to a plea for inquiry at the "meso" level into "institutions intermediate between the firm and the market, or the individual and the state." Though Misa asserts that working on this plane will offer "a framework for integrating the social shaping of technology and the technological shaping of society," it is far from clear whether this augurs the fashioning of a new, more complex and inclusive master narrative or a collage of intersecting, conflicting, and hence ambiguous tales. In either case, Misa remains vague about *how* intermediate institutions process inchoate "micro" activities into the overarching rationalities of macro patterns (if they do so). This is but to say that the meso level of analysis is at present undertheorized, as well as far too rarely investigated empirically.

12. Geoffrey Hodgson, *Economics and Institutions* (University of Pennsylvania Press, 1988), pp. 108–114.

market relations.[13] Given the fact that these quantifications have failed to be persuasive beyond policy circles, one could also ask why historians should accept such transformations of qualitative factors into highly conjectural numerical computations. Denied the tidy successions of technical progress, the transcendence of technological determinism, and the quasi-science of cliometric reduction, yet alert to context and conflict, historians of technology must confront the decentering, if not the death, of progress as a conceptual pivot for research. What is to be done?

From the preceding song, it should be obvious that I work from a "contextualist" perspective, not an "internalist" one. The way in which I chose, speculatively and elliptically, to address the notion of technological determinism in this essay's opening paragraphs tied it to place and period, to policy and contest. This may be exasperating to some colleagues; certainly the above sketch would be rejected by scores of historians in other fields for its insufficient attention to culture, markets, gender, the labor process, or the diversity of individual and institutional appropriations of technology, and/or for its elision of power relations. Whether that sketch is defensible is of minor interest here. Of greater salience is the fact that its defects speak to a second-order decentering. History as past politics is moldering. History as national saga or the consequence of relations between economic base and socio-cultural superstructure seems laughable. Historical study has devolved from period to field to multiplying concepts, methods, and theories, profoundly undercutting efforts at synthesis, preservation of a canonical narrative, or the naive "addition" of new knowledge to received wisdom. Incommensurable approaches carve out fiefdoms in the imaginary terrain of subdisciplines, even as the cry for unity and integration echoes.

Though it has identified august forebears, as a field the history of technology is a relative novelty. My sense of its course, drawn largely from John Staudenmaier,[14] suggests a movement from narrow instance-and-artifact-centered studies toward wide-ranging engagement with what Charles Tilly denominates as "big structures, large

13. Michiel Schwarz and Michael Thompson, *Divided We Stand: Redefining Politics, Technology and Social Choice* (University of Pennsylvania Press, 1990), chapter 7.

14. John Staudenmaier, *Technology's Storytellers* (MIT Press, 1985).

processes, and huge comparisons,"[15] ultimately reaching for full admission to the Castle of History. Yet, after a generation of hammering, the gates have opened to reveal a ruin. A few barons have locked themselves in the tower, but all else is in disarray. There has been no orderly, ritualized, Golden Bough-like dispatching of the elders, but a shattering of the walls of the edifice outside which historians of technology labored for recognition. Quantifiers, analysts of gender, deconstructionists, neo-Marxists, post-structuralists, and others have reduced the castellated battlements to rubble. No center has held or stands to be beheld.

Although occasional arrows still take flight from the barons' keep, the ravagers are beyond their reach or armored against their effect.[16] In essence, the circle we sought to join has broken into fragments, not so much warring as indifferent factions. The collective project we sought to infuse with our insights and energy looks to be an exercise in nostalgia. One response to this situation might be to tend such gardens as we have marked off, stepping back from the mess. As ever, denial is available; adopting a "this too shall pass" attitude has its charms. Yet if we take seriously the dispersion of history, along with the vigor of our fellow practitioners, a bundle of mid-range and manageable challenges can be specified.

Clearly the history of technology faces an environment of serial indeterminacies. As the essays in this volume suggest, there is a radical openness on questions of epistemology, method, significance, rhetoric, inference, and evidence that could augur disciplinary incoherence. Given this, I shall offer three general notions, four propositions, and three specific issues for consideration, the last group drawn explicitly from my own work. They surely range from the benign to the dubious, but broaching them may be provocative at worst and of practical utility at best. In no way, however, can they be projected as an effort to constitute a new core problematic.

15. Charles Tilly, *Big Structures, Large Processes, Huge Comparisons* (Russell Sage Foundation, 1984).

16. This image was suggested by how lightly Gertrude Himmelfarb's blasts against new historical forms were taken at the plenary session of the 1988 American Historical Association convention in Cincinnati. See "AHR Forum: The Old History and the New," *American Historical Review* 94 (1989): 654–698.

First on the ledger of general notions: If a grand technological determinism is implausible, might there not be specific locales where, and temporal envelopes when, something like Hughes's technological momentum is sufficiently powerful to overcome the constraints offered by other factors in the situational context? This is to suggest that, whereas technical advances and the quest for innovation and efficiency are not universally regnant, there may well be sites, sectors, and periods in which a technology-oriented logic governs. Such a concept of local determination (rather than comprehensive determinism) calls for specification and differentiation.[17] How does the lust for technical novelty overcome the calculus of risks, the conservatism of traditional practice, and the fear of failure? To what degree are non-market and institutional forces enabling? How are such possibilities "mapped" across sectors, from foundries to cigarettes and rice culture? What may be their implications for the labor process, market structure, sectoral governance, and the definition and exercise of power within firms and the larger political economy?[18] At base, what makes for technical dominance, and why is it not found more often? One exemplary response to several of these queries will be noted shortly.

One means for identifying such local determination has been suggested by Hughes's investigations of technology-dependent systems, especially in power generation and military weaponry.[19] The effects of bureaucratic practice and organizational drag in depleting technological momentum are stressed in his study, but the unit of analysis and criteria for systematicity outside these spheres are not at all

17. Richard Walker, "Technological Determination and Determinism: Industrial Growth and Location," in *High Technology, Space and Society,* ed. M. Castells (Sage, 1985).

18. Here Alfred Chandler's assessment of the technical differences between throughput-capable and non-throughput sectors (*The Visible Hand* (Harvard University Press, 1977)) is suggestive. "Sectoral governance" refers both to the internal organization of firms and to the extra-firm activity of formal and informal institutions, such as trade associations, research and technical institutes, lobbying groups, clubs, and collective marketing agencies. See Michael Storper and Bennett Harrison, "Flexibility, Hierarchy, and Regional Development: The Changing Structure of Industrial Production Systems and Their Forms of Governance," Working Paper D902, UCLA Graduate School of Architecture and Urban Planning, 1990.

19. Hughes, *American Genesis.*

obvious. We need to think through in what sense the DuPont corporation,[20] the auto or the leather industry, and Silicon Valley were or are systems in anything like the same sense.[21] How is it that the drive to technical innovation falters *outside* contexts of bureaucracy and regulation? Where may we find examples of its reassertion and renewal, rather than occasions of dissipation and system maintenance after the first burst of technical advance?[22] The rigidities Hughes documents as constraining innovations that would unsettle established relations leave room for outsiders, such as the garage engineer who picks up a nineteenth-century mantle. What conjunctures of locale, sector, and period combine to foster this sort of independent exploration, especially in the current century, and what social, political, and institutional mechanisms sustain it? Such inquiries could also help the conceptualization and documentation of complementarities and antagonisms between systemic and extensive patterns of technical change.

These thoughts lead into the second general consideration: What do we mean by "context"? If it is inadequate to view technology as an autonomous, unmoved mover, and if technical change has meaning and influence only situationally, how may we best specify these settings and the relationships they embody? Often enough, I think, context is treated either theatrically or structurally. In the first case, the environment is constituted as a static stage set (proscenium, curtains and backdrops, bits of furniture) on which the action takes shape. In the second, it surfaces as a network of constraints operating to condition, obstruct, or deflect the initiatives of individuals and institutions. In the theatrical frame, agents' efforts affect the set little; their implications affect later actions, for which the backdrops appear magically. A structural context leads to rhetorics of breakthrough and heroic overcoming of barriers (the Manhattan Project?) which

20. David Hounshell and John K. Smith, *Science and Corporate Strategy* (Cambridge University Press, 1988).

21. Annalee Saxenian, "The Urban Contradictions of Silicon Valley," *International Journal of Urban and Regional Research* 7 (1983): 237–261; "In Search of Power: the Organization of Business in Silicon Valley and Route 128," *Economy and Society* 18 (1989): 25–70.

22. Thomas Peters and Robert Waterman (*In Search of Excellence* (Harper and Row, 1982)) offer 3M and other corporations as recent examples, but this is of little direct value to historians of technology.

do change the environment, sometimes in unanticipated ways. Yet neither engages context itself as a process.

A familiar phrase may help create a fruitful analogy to help develop this point: Individuals often complain that something they said or wrote has been "quoted out of context." Generally this indicates that a "sound bite" has been extracted from the surrounding text, distorting a meaning that would be evident were it restored to the original sentence, paragraph, or article. What is at stake here is a struggle for control over meaning, a process of constituting contexts, with the author and user each making different claims about the proper frame within which to gauge significance (consider political campaign ads). Our activity as historians involves just this process of constitution; we author contexts for establishing meaning that are *de facto* distinct from those occupied by the actors and phenomena we represent (not re-present). We do not simply *find* the right theater in which to mount our dramas; instead we *build* it. In so doing, we make (usually implicit) claims that our construction is appropriate, that the things we left out are irrelevant, and so forth. As we proceed, we might well keep two things in mind. First, structural features of an environment (e.g. institutions) possess both constraining and enabling capacities, which materialize only in social practice (a matter Anthony Giddens has emphasized[23]). Second, as Mark Granovetter has noted,[24] action does not happen on the surface of context; action and context are mutually and intricately embedded in one another. Individuals act both in and upon institutions (a family, a firm, a government), and they embody, realize, and reproduce those institutions through their daily activity. Such a perspective does not reduce technological change to a dependent variable within some other autonomous process (capitalist development, state formation), but rather entwines it with social practice and invites us to specify those historical dynamics to which it is critically salient and to assess others to which it is arguably peripheral (the enclosure movement? public education?). As historical practice, it leads away from expositions located within scientistic constructs of social action and toward J. H. Hexter's explanation-why narrative accounts (validated by the

23. Anthony Giddens, *The Constitution of Society* (University of California Press, 1984).

24. Mark Granovetter, "Economic Action and Social Structure: The Problem of Embeddedness," *American Journal of Sociology* 91 (1985): 481–510.

reader) rather than toward method accounts (which are not the "property" of authors, but which are instead subject to mutable use by all who encounter them).[25]

Here it may be apposite to shift from the abstract to the empirical for a moment, and to examine a site at which local determination and technological change intersected. In the last quarter of the nineteenth century, the vital center of American industrial development shifted from the Atlantic coast to midwestern cities, large (Cleveland, Cincinnati, Chicago) and small (Grand Rapids, Rockford, South Bend). Parallel with this surge came a demand for machinery, particularly machine tools for metal- and woodworking. Though Cleveland boasted the innovative Warner and Swazey firm and Chicago had the region's fastest-growing and most complex industrial structure, it was at Cincinnati that technological innovation in tool building rooted and expanded dramatically, turning the city and nearby towns into the nation's most respected machine-tool district well before the First World War.

On the face of it, this is a curious outcome, for Cincinnati lay off the main east-west railway corridors, its river traffic and related activities were waning, and its principal trades had arisen from bulk processing of agricultural materials (slaughtering, distilling, brewing, and the manufacture of tobacco goods). To be sure, Cincinnati's leading enterprisers sponsored mechanics institutes and industrial expositions, but this was the case elsewhere. Here is where local agency and the structuring of context mattered a great deal: Cincinnati, like other river and lakeside sites, had had at the middle of the nineteenth century a modest steamboat-building sector, which had decayed steadily as railway capacities advanced. However, riverboat engine-building metalworkers had rapidly turned their attentions to machinery making, first in the general machine trade (still common in the East). By the 1870s, they had shifted their focus to tools for cutting wood and metal.[26] First from the shops of John Steptoe, then

25. J. H. Hexter, *The History Primer* (Basic Books, 1971), chapters 5–7.

26. It should be noted that tools for metal-forming (presses, forges, etc) were never much developed in Cincinnati—in large part, I suspect, because of the mixed wood- and metal-oriented paths of the early local tool builders. Though there were some presses developed elsewhere for embossing wood chair parts, more intensely metal-centered districts (Buffalo, Cleveland, Philadelphia, Erie, Tiffin, Chambersburg) spawned tool-building firms devoted to forming. See *American Drop Forger* 8 (January 1922): 15–26.

more forcefully from those of William Lodge (a Steptoe worker turned entrepreneur), dozens of skilled operatives and foremen stepped out to start their own machine-tool manufactories, initially sustained by elaborate networks of subcontracting from older firms. Relations of shop-floor mutuality were reconfigured into fraternal networks among novice and veteran proprietors that created a context for decisive technological change. In the 1880s three structuring patterns took shape, following Lodge's lead. More and more new firms centered their efforts on one line on tools (lathes, planers, millers)—a tactic that, in an environment of readily shared information, focused technical problem solving rather than spreading it across an indefinite range of machines. Second, shops sequentially refined designs for different classes within their specialties so as to use batch-produced components in a variety of models; this had both economic and engineering benefits. Third, a collective marketing scheme was devised through which a single agent secured orders for many firms' complementary products and then channeled them to the 20 or more shops along Mill Creek.[27]

This evolving tapestry of context-framing actions and action-inducing contexts generated a technological momentum different from that to which Hughes refers,[28] for it was open, non-systemic, locally determined, and as much dependent on social relations as on engineering refinements. As with present-day high-technology districts, its vitality demanded regular renewal (long supplied by migrations of skilled workers toward the sectoral magnet, plus institutionalized training in the University of Cincinnati's cooperative engineering course) rather than being sustained and routinized by generalized investments in transportation, power, or military infrastructures. That there may be varied configurations of technological momentum is an implication of this example that is perhaps worth further exploration.

These ruminations broach the third general concern: the importance of struggling with theory, an effort Heilbroner largely rejects

27. George Wing, The History of the Cincinnati Machine Tool Industry, DBA dissertation, Indiana University, 1964; Joseph Roe, *English and American Tool Builders* (Yale University Press, 1916); Philip Scranton, "Endless Novelty: Flexible Production and American Industrialization, 1870–1940," unpublished manuscript.

28. See Hughes's essay in this volume.

(despite Williams's incisive prodding) and Smith and Misa embrace in rather different ways in this collection.[29] My invocation on behalf of theory has two parts, one banal and the other problematic. We historians are often reminded that, in addition to mastering our trade's literature and digesting mounds of sources, we must be alert to and explicit about our assumptions, interrogate them in the process of research and writing, and, less often, expose their links to theoretical stances. Blessed and cursed by our use of ordinary formal prose, we often rely silently on theoretical expectations about modernization, bureaucracy, rational choice, or market dynamics drawn from sister disciplines. All too rarely are we fully conversant with their attendant commitments and limitations. Being criticized for insufficient theoretical understanding can push us back to the garden of empiricism and "simple" narrative. The alternative is not to "buy into" some theoretical line (neoclassical economics, or some version of Marxism or deconstruction), but to wade into the multiple streams of current discourse and struggle to make useful sense of their angles of approach.

This opens the door to the problematic part. In my experience, reading theory demands dialogue; yet history is with few exceptions the province of the solitary, even lonely, scholar.[30] Shared commitments to reading groups or seminars—which some institutions actively support, seeking cross-discipline participants—can yield rich benefits for research and teaching, along with durable collegial links. Theory and practice are interactive and, like Ruskins's thought and labor, "cannot be separated with impunity."[31] Provocative perspectives can be found at the level of mid-range or field theory in the work of David Harvey, Michael Storper, Scott Lash, Richard Urry, Edward Soja, and Richard Walker in geography; Joan Scott, Evan Connell, Teresa de Lauretis, and Linda Nicholson on gender; or Fred Block, Oliver Williamson, Amitai Etzioni, Geoffrey Hodgson,

29. Smith draws fruitfully on critiques of post-modernism and post-structuralism, whereas Misa engages the limits and variations of rationality literature.

30. George Kennan, "The Experience of Writing History," in *The Vital Past,* ed. S. Vaughn (University of Georgia Press, 1985), pp. 92–95.

31. The Ruskin phrase comes from a poster which hangs in my study. Its textual source is unknown to me and is not listed in *Bartlett's Familiar Quotations.*

and Nathan Rosenberg in economics—to say nothing of Clifford Geertz, Marshall Sahlins, and others in the social sciences. For grand or general theory, one can visit with Derrida, Foucault, Habermas, Elster, Gadamer, Feyerabend, Rorty, or Giddens, all of whom (as Quentin Skinner has noted) "emphasize the importance of the local and the contingent."[32] Whereas history-of-technology studies concerning symbol and gender have used segments of this theoretical work imaginatively,[33] elsewhere it has been appropriated mechani-

32. Such a recitation is obviously prescriptive and incomplete. Nevertheless, representative works by those mentioned include the following: David Harvey, *The Urbanization of Capital* (Johns Hopkins University Press, 1985); *The Condition of Post-Modernity* (Stanford University Press, 1989); Michael Storper and Richard Walker, *The Capitalist Imperative* (Blackwell, 1989); Scott Lash and Richard Urry, *The End of Organized Capitalism* (University of Wisconsin Press, 1987); Joan Scott, *Gender and the Politics of History* (Columbia University Press, 1988); Evan Connell, *Gender and Power* (Stanford University Press, 1987); Theresa de Lauretis, ed., *Feminist Studies/Critical Studies* (Indiana University Press, 1986); Linda Nicholson, *Gender and History* (Columbia University Press, 1986); Fred Block, *Postindustrial Possibilities* (University of California Press, 1990); Oliver Williamson, *The Economic Institutions of Capitalism* (Free Press, 1985); Geoffrey Hodgson, *Economics and Institutions* (University of Pennsylvania Press, 1988); Amitai Etzioni, *The Moral Dimension* (Free Press, 1988); Nathan Rosenberg, *Inside the Black Box* (Cambridge University Press, 1982); Clifford Geertz, *Local Knowledge* (Basic Books, 1983); Marshall Sahlins, *Culture and Practical Reason* (University of Chicago Press, 1983). Comparable rosters of work in state theory, neo-Marxism, or cultural theory could be assembled without difficulty.

Sketches of most of the general theories and theorists mentioned may be found in Quentin Skinner, *The Return of Grand Theory in the Human Sciences* (Cambridge University Press, 1985) (quote from p. 12). For others not included in Skinner and noted here, see Richard Rorty, *Philosophy and the Mirror of Nature* (Princeton University Press, 1979); Paul Feyerabend, *Farewell to Reason* (Verso, 1987); Jon Elster, *Explaining Technical Change* (Cambridge University Press, 1983); *Nuts and Bolts for the Social Sciences* (Cambridge University Press, 1989); *The Cement of Society* (Cambridge University Press, 1989); Anthony Giddens, *Central Problems in Social Theory* (University of California Press, 1979); *The Constitution of Society* (University of California Press, 1984); *The Nation State and Violence* (University of California Press, 1987).

33. See Rosalind Williams's *Notes on the Underground* (MIT Press, 1990) and her essay in this volume for evidence of an effective meshing of theoretical insights with the rhetoric of exposition. Rather than setting theory off in a preface, Williams infuses the text with relations to multiple theoretical streams (Adorno, Raymond Williams, Derrida, Barthes, Durkheim, Eliade).

cally or simply left aside. On each of these three counts—local determination, articulating context as process, and approaching theory—there is indeed much worth doing. Such efforts can usefully inform each of the thematic areas of empirical work that Staudenmaier has identified.[34]

Four observations or propositions based loosely on the foregoing may merit attention. First, if we set totalizing determinism aside, we abandon any notion that shifts in technology govern the restructuring of social formations (families, schools, firms, governments) or of cultural practices. Just as revisionist Marxism has grappled with the implausibility of "reading off" social relations from changes in the forces of production (in part through reworking Gramsci's notion of hegemony),[35] historians of technology need to conceptualize the multiple oppositions and disjunctures that surface in eras of swift technical change without relying on overarching teleologies. The problem here is how to capture the dynamics of interacting intentionalities involving power, resistance, and consent while steering clear of fresh reductionisms that merely substitute a new universal (bureaucratization, the market) for the contingent play of historical agency.

Further, any assumption that technical change represents a unified process, explicable perhaps by reference to leading and lagging sectors, is hardly sustainable. Such a framework directly entails the presumption that diversity is of little moment. Surely we could identify six or a dozen distinct currents of technological change unevenly mapped across period and place, some complementary to and some disruptive of others. The pattern of their interaction effects and their unintended consequences will defy any linear model, any ordering by factor analysis. Culturally, such complexity invites us to consider the difference between technologies' ascribed significance in historical conjunctures and our retrodictive valuations, now that we presume to understand which were the "winning weapons"—a point brought out tellingly in Thomas Misa's essay in this volume. Moreover, as Ruth Cowan's vivid treatment of the gas refrigerator suggests,[36] the exploration of technological alternatives beckons as well. If the convergences on the QWERTY keyboard and the electric "fridge" were

34. Staudenmaier, "Recent Trends."

35. For a persuasive reading see Butsch, *For Fun and Profit,* pp. 3–19.

36. Ruth Swartz Cowan, *More Work for Mother* (Basic Books, 1983), chapter 5.

not due to the internal merits or the efficiencies of these technologies, then what conditions situations in which diverse technologies coexist (as in foundry practice, or as in cost accounting)?[37]

Third, in view of the above, the situational links between technical shifts and social and political relations have too readily been left unspecified and underinvestigated. Specifically, within a determinism, in order to operate as a primal force that alters other social phenomena, technical change must be insulated from being viewed as a *consequence* of extra-technical initiatives, as Perdue has rightly stressed.[38] Much follows from this autonomy; however, if it is relativized or jettisoned, the temptation to locate such a force in implicitly autonomous organizations can be magnetic. Such a recoding, fetishizing the firm (or the state), evades basic questions about the heterogeneity of business interests and means to advance them, about the boundaries of the firm, about institutional authority, and about the social, cultural, and political values that condition the specification of choices and enable or constrain decisions.[39] Even a situational "technical logic" will depend on the forms of rationality (e.g., moral vs. instrumental) that are involved in contests over change, and on the social constructions of gender, cognition, and knowledge that validate modes of argumentation and criteria for bounding decision spaces.[40] The firm, the research institute, and the military proving ground are sites of plural rationalities, rival sets of rules for closure and models of consent—in essence, domains of indeterminacy (or the "inchoate," as Schwartz and Thompson put it[41]) rather than handy black boxes for technologies. One implication of this line of thinking is that there

37. Howell Harris, "Little Drops of Water, Little Grains of Sand," unpublished manuscript, University of Durham, 1991; H. Thomas Johnson and Robert Kaplan, *Relevance Lost: The Rise and Fall of Managerial Accounting* (Harvard Business School Press, 1987).

38. See Perdue's essay below.

39. For provocative perspectives on these matters see Sharon Zukin and Paul DiMaggio, *Structures of Capital: The Social Organization of the Economy* (Cambridge University Press, 1990).

40. Carol Gilligan, *In a Different Voice* (Harvard University Press, 1982); Scott, *Gender;* Hodgson, *Economics and Institutions;* Schwartz and Thompson, *Divided We Stand,* chapters 1–3.

41. Schwartz and Thompson, *Divided,* pp. 6–13, 131–135. On multiple rationalities see also Burton Klein, *Dynamic Economics* (Harvard University Press, 1977).

may well be no way to establish reliable dependent-independent variable sets (such as are common in social science) for use in a contextualized history of technology, again making forthrightly interpretive, theoretically explicit narratives a plausible format for research practice.[42]

One example of a means to approach technology as process might be built upon Joan Scott's essay "Gender: A Useful Category of Historical Analysis." In addition to its detailed review of literatures and modes of addressing gender questions, Scott's article also offers a methodological claim that her four-part scheme for studying gender relationships "could be used to discuss class, race, ethnicity, or for that matter, any social process." The levels of analysis for interpretation include "culturally available symbols that evoke multiple (and often contradictory) representations," "normative concepts" that fix dominant meanings of such symbols, the institutional complexes in which these are articulated, and the way in which the relevant symbols, norms, and institutions are implicated in the construction of subjective identities. Without undue stretching, this framework could inform varied historical studies of technology—especially if careful effort is directed, as Scott suggests, to establishing "what the relationships among the four aspects are."[43] Moreover, though doubtless more restricted than gender effects, it is reasonable to seek how, when "established as an objective set of references," concepts of technology "structure perception and the concrete and symbolic organization of all social life," as Scott says concepts of gender do. "To the extent that these references establish distributions of power (differential control over access to material and symbolic resources)," Scott continues, gender "becomes implicated in the conception of construction of power itself"[44]; I would say the same of technology. Thus power—often pushed to the margins of efficiency centered accounts of technical change—would be reintegrated into a pro-

42. Schwartz and Thompson's critique of technological risk assessment drives home the deeply problematic character of social science methods and epistemological claims (*Divided,* chapter 7). For a sustained discussion of "post-empiricist" epistemology see Paul Roth, *Meaning and Method in the Social Sciences* (Cornell University Press, 1987).

43. Scott, *Gender,* pp. 43–44.

44. Ibid., p. 45.

gram of research, one vector of which Michael Smith's essay above represents.

Fourth, once we move beyond linear and reductionist accounts of technological change, we will notice a variety of silences and great forgettings in the history of technology. Staudenmaier has marked out non-Western concepts of technology and technical practice, as well as technology-environment relations, as signal examples. Others might point to technologies of sexuality and family limitation, or to technologies for the "management" of the incarcerated or the dead. Thanks to Linda Gordon, Michel Foucault, and others, galleries such as these have been hewn open, none of them derivative of the grand trajectory from the spinning frame to the nuclear reactor.[45] Within the industrial realm, further silences and blocked-off avenues may be noted. In my own work, these are found in the world of batch and custom production, which somehow fell off the map of industrial history as scholars pursued the origins and mutations of mass production in America. It was not their practices but our conceptualizations that designated shipbuilding, furniture making, printing, and foundry casting as backward or peripheral. Indeed, once we reconstitute the shop technologies and relations along with the managerial, marketing, and associational techniques of these specialty enterprises, we discover a different construction of industrial prowess. A distinctive factory culture, with its own requirements for and patterns of technical practice, developed alongside and in interaction with the mass-production format. Despite its various limitations, it represented an environment of versatility, flexibility, and (not infrequently) mutuality—attributes eagerly championed in present-day business circles. In revisiting such overlooked yet heavily traveled paths, we may identify both the situational elements that fostered the vitality of batch production and the institutions and processes that conditioned its uneven decay in the first half of the twentieth century.[46]

In this light, three empirical issues that arise from my own research efforts are worth highlighting. Though considerable work has been devoted recently to appreciating "difference" and its social articula-

45. Staudenmaier, "Recent Trends," p. 725; Linda Gordon, *Women's Body, Women's Right* (Penguin, 1983); Michel Foucault, *Discipline and Punish* (Random House, 1977).

46. Philip Scranton, "Diversity in Diversity: Flexible Production and American Industrialization, 1880–1930," *Business History Review* 65 (1991): 27–90.

tion, I am intrigued by the parallel social construction of "sameness." It has wider implications, but in regard to commercial products it may be understood as related to the economic impact of substitutability. Sameness is the notion that if two goods or services provide the same utility, they are interchangeable, with trade preference going to the provider offering a lower price (or, failing that, better terms, faster response, etc.). In the late-nineteenth-century United States there were a number of such commodities in the economy, including insurance contracts with identical coverage, iron nails of a specified size, and staple cotton print cloths of a standard width and construction. However, coal mined in different regions (even adjacent counties) had differing energy outputs, leather was far from uniform, pig iron varied in strength according to its ore bases and processing techniques, and railway service was notoriously difficult to reduce to standard measures. In simple finished goods—wool blankets, work shoes, newsprint—there was so much diversity that the notion of sameness was essentially an artifact of hope. Buying by price indicated a deficiency in information, for cheaper meant different, not substitution. (Cheapness could also indicate strategic or stupid pricing by producers, the former to increase market share—but how was a buyer to know that sameness in use qualities had been preserved? Such issues certainly fed the rise of grading systems and testing labs for materials.[47])

For fashion fabrics, furniture, or locomotives, product difference was ineluctable. Yet by the 1920s, and in some sectors earlier, price distinctions governed the sale of worsted suitings, pine tables, and the like. How was this suppression of difference achieved, and how might this achievement have advantaged enterprises increasingly devoted to cost minimization relative to those focused on ensuring quality and diversity of output? Amid such a dynamic, technological commitments that emphasized speed and routinization would have greater yield than those centered on quality of output and flexible scope, for example. Moreover, in the early twentieth century, parties other than the producer and the client were interested in sameness,

47. Susan Strasser, in *Satisfaction Guaranteed* (Basic Books, 1989), has worked to illuminate the converse: the ways in which makers of staple soaps, biscuits, etc. struggled to construct difference so as to demand premium prices for mass-produced brand-name goods. See also Roland Marchand, *Advertising the American Dream* (University of California Press, 1985), pp. 120–132.

though initially they did not appreciate it as a social construction. For example, in 1912 the Federal Bureau of Corporations undertook a study of "efficiency" among "trusts" in response to concerns about industrial concentration. The investigators examined sectors having the most standard products—such as paper—and soon found themselves in a terrible mess.[48] The manufacturers they interviewed, makers of newsprint, writing, and wrapping paper who turned out products assumed to be uniform, were quite certain that from firm to firm such goods were not the same. Each company had detailed price lists, but when they had to shade prices to get contracts they "fiddled the goods," altering their composition in various ways to preserve expected levels of profitability. The idea that there was some fundamental basis for devising a means to measure efficiency in production at firms of all scales dissolved. The study was abandoned, for the researchers were not prepared to pretend that identities existed where they had discovered differences. Yet this was a rare case. Twentieth-century America labored to construct sameness, and that construct became a weapon in economic rivalries and policy formation as well as a means through which to construe the world of goods.[49] How this happened and what its implications might be are issues of genuine interest.

A closely related topic is what might be called the contest over standardization in the United States from the 1890s through the 1930s. It is easy to view standardization as an almost naturalistic process guided by efficiency criteria. However, standardization stood at a contentious intersection of what Foucault termed the three "great forms" of liberal reason: objective thought, technical apparatus, and political organization.[50] In the battle over standardization, the state, scientists, and engineers (and instrumental reason) were all contending with the brute realities of making things and selling them. Arguing that standardization was good science and a key to modernity

48. Walter Durand, "Preliminary Memorandum on Suitability of Various Branches of the Paper Industry for an Efficiency Investigation," Box 94, File 1361-3, Records of the Federal Trade Commission and Bureau of Corporations, RG 122, National Archives, Washington, D.C.

49. Mary Douglas and Baron Isherwood, *The World of Goods* (Basic Books, 1979).

50. Michel Foucault, *Foucault Live: Interviews, 1966–84* (Semiotext, 1989), p. 240.

and enhanced efficiency, visionaries and bureaucrats were surely frustrated at the resistance they met once the question "Who benefits?" was raised sharply. Large users of intermediate goods and machinery wanted their suppliers to standardize, thus reducing variety, parts inventory, and dependence; but who was to bear the costs of change, and how would suppliers benefit? To suppliers it looked to be a tactic that would intensify competition and bring lower selling prices, with only uncertain prospects for generally larger demand. Though consumers' interest in the cost gains from standardization was often chorused, difference and distinctiveness had long been selling points in scores of product lines. At what levels did the interposition of standards make sense, and for whom? The struggle peaked in the years after World War I, sparked by a much-publicized concern with "waste in industry"; but the very definition of waste was highly politicized, with powerful institutions and interests favoring and opposing standards for economic and cultural reasons (witness the repeated rejection of the metric system).[51]

These matters can be related productively to my four earlier propositions: (1) that technological change proceeds in the absence of overarching rationalities, (2) that it proceeds along multiple coexistent trajectories, (3) that links between technical change and sociopolitical relations are intimate and underspecified, and (4) that stepping beyond reductionist teleologies reveals a array of intriguing silences in the history of technology. Issues of product variety and the construction of sameness exemplify the conflict between diverse forms of industrial performance and a conjunctural ideological campaign to define efficiency. They involve both the state and professionalizing institutions in propagandizing standardization, and they reframe an earlier marginal story of progressive visionaries and ill-informed laggards as a far more complex and significant struggle between rival visions of industrial prowess. That an inclusive and critical history of standardization in the United States has yet to be researched and written underscores the enduring, wrong-headed assumption that this was a non-problem.

51. Among the best sources for this process are publications of the Department of Commerce's Division of Simplified Practice and frequent articles in Shaw Syndicate business journals (*Factory, System*) in the 1920s. For a brief discussion of the Hoover Administration's efforts see William Barber, *From New Era to New Deal: Herbert Hoover, the Economists, and American Economic Policy, 1921–33* (Cambridge University Press, 1985), pp. 13–15.

A second issue that my research on batch production has raised has to do with how we view the development of machinery. It seems to me that we have often studied machines as though there existed a single vector of advance toward the mass-production technologies that defined the American System. We have examined closely the fast and the simplified, the machine as an agent of the division of labor, leading to dedicated technologies. But if we commence in the 1880s and survey machinery in the non-throughput industries, we will uncover lines of development targeting the enhancement or conservation of versatility. Diverse machines and appliances (some produced at the Cincinnati complex) embodied flexibilities that could be drawn on by moderately skilled workers. One did not have to be an artisanal genius to do many different tasks with such devices. Let me offer four examples I have encountered.

One of the most significant machines of the era was the papermaking Fourdrinier,[52] which could be adjusted, as one informant told federal efficiency researchers in 1912, to make hundreds of different kinds of paper, ranging from board to the finest rag, of varied colors, watermarked, laid, or plain, at the buyer's pleasure. This huge machine represented versatility structured into technology. Of course, the Fourdrinier's capacities were not always exploited, as some mills confined their operations to a small group of staple items. Yet wherever variety and quality were stressed, shorter runs were the rule and the skill dimension remained, along with the need for managerial planning for production and marketing of diverse outputs. In woodworking, the machine known as a "combination" was roughly analogous. A combination had a lathe, a planer, a rabbeter, and a number of other functions articulated together in a large, rectangular block for transportation to construction sites. From four to seven people could work on a combination simultaneously or sequentially, cutting, shaping, and sanding wooden structural and ornamental pieces.[53]

52. See Judith McGaw's *Most Wonderful Machine* (Princeton University Press, 1987) for a discussion of classic Fourdrinier operations through 1885. See the above-cited files of the Bureau of Corporations (n. 48) for a review of their capacities c. 1912.

53. For combinations, see articles and advertisements in *The Wood-Worker* (1900–1920), a weekly journal of the lumber, furniture, and construction trades.

In metal cutting, the turret lathe (which eventually became one of the major tools in the production of auto parts) was designed for versatility, as Oberlin Smith of New Jersey's Ferracute Machine Company observed in the 1890s.[54] A turret would accept a variety of cutting tools and bring them to bear on castings for a run of 30 or 50 or 50,000 parts, in batch cases shortening set-up times and changes, simplifying shop orders and work planning under conditions of product diversity. For textiles, the range of the Jacquard loom, and its knitting companion, the Raschel, are well known; each was capable of reproducing an indefinite range of patterns by "reading" punched cards or tapes.

All these devices encapsulated diversity in their physical forms, yet our understanding of the trajectories of their development is so fragmentary that expanding it could occupy a number of "internalist" historians of technology for a fair stretch. In light of my propositions, the patterns by which these devices developed were not uniform and did not follow a comprehensive logic. New models of the turret lathe succeeded one another rapidly, greater power and stability and improved lubrication being among the gains; however, a 1930 Wharton School analysis noted that improvements in the Jacquard had barely been noticeable since the turn of the century.[55] Was this difference simply a result of the divergent courses of the auto and textile sectors, or was something more complex at work—something that relates to Cohen and Zysman's suggestive notions of "dynamic" and "static" flexibility?[56] Machinery development's links to ideology, state, and society, moreover, have rarely been probed by researchers. Though they have not been systematically explored, these examples suggest that such machines supplied versatility and supported a technical virtuosity of a very different sort than historians have customarily treated.

54. Oberlin Smith, "Modern Machine Tools," *Engineering Magazine* 8 (1894–95): 54–61.

55. Industrial Research Department, University of Pennsylvania, "Final Report to the Manufacturers and Union Members of the Philadelphia Upholstery Industry," 1931, chapter 4, pp. 9, 10, Box 135, Gladys Palmer Papers, Urban Archives, Paley Library, Temple University.

56. Stephen S. Cohen and John Zysman, *Manufacturing Matters: The Myth of the Post-Industrial Economy* (1987), pp. 131–134.

Lastly, there is a labor-process dimension to research on batch and custom manufacturing, indicating that a distinct form of the "employment relation"[57] may have characterized much of flexible manufacturing. As Fred Block explained in comments on a paper I offered in 1989 at a University of Pennsylvania seminar,[58] economists and labor historians have usually regarded work relations as being either contractual or coercive. "Contractual" refers to the contours of labor markets as well as to definite agreements on terms, conditions, and expectations for employees, whereas "coercive" refers both to structured technologies under unilateral management control and to market-failure settings in which free and fair contracting is severely constrained. Yet my research kept uncovering workplaces in which there appeared to be a spirit of cooperation, with problems solved in collaborative ways—plants where there was evidence of reciprocity between workers and owners or managers. Though recognition of this third-stream "cooperative impulse"[59] can readily lead to nostalgia for an imagined Golden Age, such relations did deliver profits to enterprises of various scales in multiple industrial sectors, often in trades slighted by historians' studies of the American System. The cooperative impulse is located at the heart of intersecting indeterminacies. It is based on the need for repeated problem framing and solving, the value of craft knowledge and technical experience, and the salience of respect and trust to quality results.[60] In such environments, factory owners realized that in buying new machinery it was essential to consult with workers, for they were intimately familiar with the capacities and the shortcomings of the present plant facilities. Maintaining such relations involved reciprocity and a rough version of technological democracy.

Such enterprises were far from harmonious. With authority and responsibility situated in the fluid relations among particular persons, conflict was unavoidable. Yet an "ethic of craft work" allowed disputes to be "negotiated within a framework of cooperation in which each

57. The term is derived from Michael Burawoy, *The Politics of Production* (Verso, 1985) as extended by Storper and Walker (*Capitalist Imperative,* chapter 6, especially pp. 166–175).

58. Philip Scranton, "The Politics of Production" (unpublished).

59. Block, *Postindustrial Possibilities,* pp. 82–83.

60. For a set of recent studies, see Diego Gambetta, ed., *Trust: Making and Breaking Cooperative Relations* (Blackwell, 1988).

side recognized the validity of the other's claims."[61] The messiness and the frequent personalism of such relations fit poorly with the reigning notions of efficiency. Workers had power in production and had to be relied upon to use it wisely, else the future of the firm (and their employment prospects) would be damaged. Seemingly petty conflicts could escalate drastically if they were perceived as betrayals of trust and mutuality. It was such contingencies and indeterminacies that F. W. Taylor's passion for order aimed to eradicate, and it was our enduring, oft-muted fascination for them that made *The Soul of a New Machine* compelling.[62]

Now that we have lost much of our capacity to articulate and reproduce collaborative labor processes, it is not surprising to find them celebrated wherever they are encountered nowadays.[63] Here again, in relation to the earlier propositions, there is value in exploring the silence that has enveloped variations in practice and in trying to understand how labor organizing (and later, state-authorized institutionalization of "one best way" for labor relations) impinged on collaborative productive relations. With a wider vision, we can document a range of effective labor-management dynamics dependent on sector, locale, and period, and we can begin better to appreciate yet another dimension of American workers' long reluctance to build and sustain unions.[64] From my angle, the imperative is to accomplish the research that will establish, first, the extent and significance of collaborative employment relations from the Civil War through the 1930s, together with the situational and institutional factors that facilitated them, and, second, the processes through which the erosion and marginalization of these relations was effected.

In closing, let me return to the challenge created by shedding determinism and acknowledging indeterminacy in the history of tech-

61. Block, *Postindustrial,* p. 83. Block likens the ethic of craft work to contemporary ethics of professionalism.

62. Daniel Nelson, *Managers and Workers* (University of Wisconsin Press, 1975); *Frederick W. Taylor and the Rise of Scientific Management* (University of Wisconsin Press, 1980); Tracy Kidder, *The Soul of a New Machine* (Little, Brown, 1981).

63. Peters and Waterman, *Excellence;* Holland, *When the Machine Stopped.*

64. This in no way diminishes the importance of employers' hostility, which has been concisely detailed by Sanford Jacoby ("American Exceptionalism Revisited: The Importance of Management," in *From Masters to Managers,* ed. S. Jacoby (Columbia University Press, 1991).

nology by way of a comment Michel Foucault offered in one of the last interviews before his death. If we are to take seriously the unhinging of determinisms that has materialized over the last two generations, we might do well to ponder Foucault's claim that "nothing is fundamental. . . . [There] are no fundamental phenomena. There are only reciprocal relations, and the perpetual gaps between intentions in relation to one another."[65] No revised socio-cultural, economic, or technological teleology is feasible. Instead, more modest efforts to unravel conjunctural complexities replete with productive complementarities and dispiriting antagonisms[66] may provide the best venues for shrugging off old myths and authoring new ones for our time—something that has ever been the historian's task.

65. *Foucault Live,* p. 267.

66. Which in this volume include the empirically rich "equilibrium" approach Perdue offers for agrarian societies, the politico-cultural interpretive line advanced by Williams and Smith, and Misa's concern for meso-level institutional analysis. It is hard, at this stage, to imagine a refined convergence among these that would yield a new and integrated framework for studies in the history of technology and society, though all of them emphasize the fullest possible reconstruction of contexts for the structuring of action.

Technological Determinism in Agrarian Societies

Peter C. Perdue

Peter Perdue turns to the comparative history of agrarian societies to "illuminate the more general question of technological determinism." Specifically, he compares Lynn White's classic "single-factor" model of technology-driven change in medieval Western European agriculture with the multi-factor "equilibrium models" of Michael Confino (on eighteenth-century Russian agriculture) and Philip Huang (on Chinese agriculture since the fourteenth century). Perdue agrees with White's critics that tightly coupled technological explanations of socioeconomic change cannot account for all the variances that have been found in the historical record. Technology constrains, he argues; it does not determine. Far more persuasive, in his view, are contextual accounts that integrate environmental, technological, social, and cultural elements. He concludes that "the long-term development of different systems does not follow any one simple model impelled by a single logic." Rather, "it is the interrelationship of all the elements, not any single one, that determines the whole."

Most discussions of technological determinism focus on industrial societies and limit their temporal scope to the last hundred years. We should broaden our view of this question by examining pre-industrial societies over several centuries. The contemporary discussion about how machines determine the evolution of human societies arises from the immense impact of modern industrial technology, whether powered by steam, electricity, or nuclear reactions. In the pre-industrial age, by contrast, the vast majority of humans worked in agriculture, using human, animal, or natural sources of power. They, too, participated in processes which they neither fully understood nor totally controlled, and they reflected on the extent to which agrarian production could be consciously directed by human intelligence. Some argue that our present ability to master the forces of nature is far greater than in the past; others argue that, even though we command enormous energy supplies, we are no more skilled than our ancestors in directing these forces toward the greatest possible human freedom and welfare. Have we progressed or not? We cannot seriously investigate such a question without closely examining agrarian societies over the long term.

In this essay I focus on technological determinism in action, as concrete interpretation and as a political program. First, I distinguish several variations of the technological determinism thesis that commonly appear in discussion of agrarian societies. Second, I demonstrate that parallel forms of this thesis reappear in widely disparate historical and geographical contexts. Finally, I explain the recurrence of these paradigms in political terms: each form of the thesis favors certain social groups, reconfirms certain authority relationships, and points to certain kinds of reforms. I discuss technological determinism both as a shaper of agrarian society and as a theory about society expounded by contemporary literate elites. Determinism acts as a theory that guides reform and as a hidden force not seen by contemporaries but discovered by later analysts.

In this discussion, I try to connect the analysis of agrarian society with the concepts used by analysts of industrial societies. If we apply the same types of argument to both, we undermine the all-too-common assumption that "real" technological change began with the Industrial Revolution and that subsequent developments followed a logic uniquely dictated by industrialism. Such an assumption is cast into doubt by recent studies of the economic history of the Industrial Revolution which explore continuities between the pre-industrial and the post-industrial age. These studies underline the importance of "proto-industrialization" and "growth recurring" (Mokyr 1985; Jones 1988). "Take-offs" are not so obvious as we once thought.

An examination of technological determinism in agrarian societies clarifies issues often obscured in present-day industrial societies. Agrarian societies do not stagnate; they develop and change like ours, but at a slower rate. Slow change becomes manifest over decades or centuries, and over such periods we can distinguish long-term trends from cyclical processes and random shocks. In the modern world, the full implications of one technology have hardly begun to play themselves out before a new technology appears: automobiles came in before the effects of the railroad had spread worldwide, and airplanes soon after. We can better untangle specific technological influences when the new does not overlap the newer.

Many analysts have assumed that technological determinism must be much stronger in agrarian societies than in industrial ones. Farmers limited to human and animal strength, supplemented by ingenious diversion of water and fire, control nature only weakly. Their vulnerability to unpredictable natural disasters—typhoons, droughts, earthquakes, floods, diseases—might encourage fatalism. Village life limits one's intellectual horizons and personal contacts; poor transportation prevents the rapid exchange of goods and ideas. For Marx, such considerations reinforced the classic stereotype of peasant life as bound by tradition, hostile to change, and sunk in routine; progress could come only after the bourgeoisie had "rescued a considerable part of the population from the idiocy of rural life" (*Communist Manifesto,* p. 339 in Tucker 1972). The peasant farmer has been the bane of urban reformers and revolutionaries for centuries.

The more historians and social scientists examine peasant societies around the world, the more the stereotypical dull peasant disappears (Schultz 1964). Primitive tools do not bind peasants to a stagnant, poverty-stricken mode of production. Under appropriate conditions, peasants do transform their societies. They adopt new methods of crop production, new seeds, new field formations, or even leave the farm, given the right incentives. If the implacable logic of technological imperatives always doomed agrarian life to backwardness, we, like Marx, would be perfectly justified in finding change only in the industrial world. The peasant could escape his misery only after the urban outlook had completely penetrated the countryside. Modernizing intellectuals—left, right, and center—have forecast the elimination of the peasant, either by sending him to work in the cities (forcibly if necessary) or by replacing him with a "farmer" indoctrinated in the modern businessman's approach to life. Urban intellectuals everywhere—Marxists, liberals, and conservatives—have fought urban battles over the future of the rural majority. Can we escape the limited perspectives of those who aim to manipulate the peasant farmer and instead explore the inner logic of agrarian life? To what extent does nature subordinate agriculture to necessity, and to what extent do peasants freely use natural resources in their own interests? In a wide variety of places and times, certain responses to these questions recur. We can illuminate the general question of technological determinism by examining these debates in several contexts.

We may consider agrarian determinism as a subcategory of environmental determinism: the assertion that the natural world, on which all humans rely for a living, fundamentally determines not only their physiological and economic well-being but also their political, social, and intellectual relationships. Herodotus, one of the first Western thinkers to espouse this idea, sharply distinguished between the civilized Greeks he knew and the barbarians of Asia. He attributed the distinction to the effects of the broad plains of Asia on the character of its inhabitants. Montesquieu elaborated on the same idea:

> Asia has always been the home of great empires; they have never subsisted in Europe. For the Asia of which we know has vaster plains than Europe; it is broken up into greater masses by the surrounding seas; and as it is further south, its springs run more easily dry, its mountains are not so covered with snow, and its rivers are lower and form lesser barriers. Power therefore must always be despotic in Asia, for if servitude were not extreme, the continent would suffer a division which the geography of the region forbids. (Montesquieu 1958, p. 529; Anderson 1979, p. 465; cf. Aron 1968, pp. 25, 35, 38–42)

For Montesquieu, climate and geography separated Eastern despotism from Western freedom. More recent East-West dichotomies also purport to link the natural environment and the political structure. Karl Wittfogel's rigid hydraulic determinism traced the need for centralized control of water supplies in oriental regimes to the inevitable development of autocratic regimes. The economic historian Eric Jones has refined the thesis in *The European Miracle: Environments, Economies and Geopolitics in the History of Europe and Asia* (1987).

Several generations of historians studying China have debunked this simplistic theory, without success. Like the *budaoweng* doll, it pops back up after every refutation. Recently, younger Chinese historians have reasserted that there is an ineradicable link between land, water, and culture. The widely broadcast TV series *River Elegy*, a brilliant combination of documentary, popular history, and academic discussion, argued forcefully that the confining impact of the brown loess soil of northern China, reinforced by the isolation created by the Great Wall, bound the minds of the Chinese people into stagnation for millennia. (See Xia 1988; Wakeman 1989.) Only the liberation brought by Western ships, sailing on the blue ocean, could bring modern ideas of democracy, scientific and technological advance, and economic progress to China. Why does *River Elegy*'s polemical, idealized history attract so many Chinese intellectuals today?

Crude versions of environmental determinism attract support when they offer simple answers to urgent problems. They often thrive on political malaise, because they remove human agency. Wittfogel's hydraulic determinism, for example, first developed in Depression-era Germany, flourished again in the postwar United States, while Communist regimes swept to power in eastern Eurasia. Activists who see no prospects of political change may console themselves by reasoning that nature determines everything anyway. This is how China looked to the producers of *River Elegy* in 1988. On the other hand, radical environmentalists invoke environmental determinism in order to galvanize change by predicting impending doom. Both distort the past in the service of reform. Don't proponents of crude determinism sacrifice science to moral vision? Is it not more useful to analyze complex interrelationships of contingency and necessity in the past, if we want to act responsibly in the future? A historian, at any rate, must believe so.

Without rejecting *in toto* the appeal of environmental determinism, some careful historians have examined both the many different constraints of the natural environment on human behavior and the human resistance to these constraints (Cronon 1983, 1991; Worster 1985; Schoppa 1989; Worster et al. 1990). They have found that the human relationship with nature faces no more inevitable constraints than any other relationship. It depends as heavily on cultural concepts (such as property rights) and on social relations as it does on land, water, and climate. Scholars of industrial science and technology have arrived at the same conclusion. The convergence of approaches to the natural, the agricultural, and the industrial environment indicates that these researchers can learn much from each other.

Three Case Studies

Let us examine these questions more concretely by turning to three very different contextual studies of the effects of technological change on agriculture: medieval Western Europe, eighteenth-century Russia, and twentieth-century China. Despite the great disparity of time and place, these studies make rather similar claims about the causes of transformation or failure to change in agrarian society. A detailed survey of the current state of scholarship on these three regions or an evaluation of the accuracy of these three works is not possible in an essay of this length; I use them only to examine efforts to link the technical means of agricultural production with social institutions.

The Heavy Plough in Medieval Europe

In chapter 2 of *Medieval Technology and Social Change,* Lynn White, Jr., following the path blazed by Marc Bloch in 1931 (see Bloch 1966), argues that the introduction of a new type of heavy plough to northern Europe allowed increases in agricultural production and in human population density. Bloch noted that the conical or triangular blade of the light scratch plough, widely used in the European lands bordering the Mediterranean, did not turn over the soil completely. It left only a shallow furrow. Farmers using it tended to make square shaped fields by ploughing a second time crosswise. This plough, however, did not suit the heavy, wet soils of northern Europe. The new heavy plough, combining a heavy vertical coulter, a horizontal ploughshare, and a curved wooden mouldboard, could turn over a large clod of earth. It thus obviated cross-ploughing. It changed the shape of fields from squares to narrow oblongs, and it allowed exploitation of fertile heavy bottom lands. Few peasants could afford the teams of eight oxen required to draw the new plough. Therefore, they had to pool their animals and labor while reallocating the village lands into unfenced "open fields"—long, narrow strips, each with a different owner, ploughed in sequence. A powerful village council determined the sequence of ploughing and decided disputes. White (1962, p. 44) concludes that these arrangements, "the essence of the manorial economy in northern Europe," are "intelligible only in terms of the heavy plough."

Building on Bloch's hypothesis, White looks for evidence of the origins of the heavy plough and finds it among the Slavs in the sixth century A.D. Spreading to Germanic areas in the seventh century, concomitant with an extraordinary increase in population density, the plough probably reached the Vikings around A.D. 800, who then took it to England and later Normandy. The effects of the plough were far-reaching. In White's words, it brought a "fundamental change in the idea of man's relation to the soil . . . once man had been part of nature; now he became her exploiter" (p. 56).

Exploitation of the plough's advantages required, however, a second new technology: the full use of the horse in agriculture. Horses use their fodder more efficiently than oxen. They pull faster and endure longer, but they are ineffective draught animals without proper harnesses and shoes. Not until the eighth or the ninth century did Europeans develop the horseshoe and the new harness, and not until the end of the eleventh century did plough horses become a common sight.

One further technological change, the evolution of the three-field rotation, completed the fully developed agricultural system of the high Middle Ages in northern Europe. Each year farmers planted one-third of their land with a winter crop (wheat or rye) and one-third with a summer crops (oats, barley, peas, or beans), and left one-third fallow. The next year, the planting was rotated, always with one field or another left fallow. The new rotation replaced the two-field rotation, dominant in southern Europe, which fallowed half the land each year. It increased the total area under cultivation, distributed the labor of ploughing and sowing more evenly through the year, guarded against famine by diversifying crops, and increased production (especially of oats, the ideal food for horses).

Thus by the thirteenth century the medieval agricultural system closely linked its three basic elements. The heavy plough supported denser rural populations who could clear more fertile land, using efficient horse traction and feeding their horses with oats from the three-field rotation. According to White, this had dramatic consequences. Peasants who ate more legumes added the complete set of amino acids to their diet, markedly improving their nutrition and vigor. This "new type of food supply . . . goes far towards explaining, for northern Europe at least, the startling expansion of population, the growth and multiplication of cities, the rise in industrial production, the outreach of commerce, and the new exuberance of spirits which enlivened that age" (White 1962, p. 76). In sum, the shift of the economic center of Europe from south to north after the ninth century, which brought a rising standard of living, a demand for manufactured goods, rapid urbanization, and the rise of classes of burghers and artisans, all goes back to this agricultural revolution of the early Middle Ages (p. 78).

The Single-Factor Method: Its Advantages and Disadvantages

Lynn White's argument epitomizes the single-factor method of investigating the influence of technology on agrarian society. White follows the influence of one isolated element, the heavy plough, on all the other elements of the system. He does not assert that this one factor inevitably determines everything else; nevertheless, he singles it out precisely because he considers it the most fundamental cause. Thus he derives a remarkable list of consequences from the introduction of the new heavy plough. His breathtaking argument traces back nearly all the significant economic features of the early modern world to the mundane technologies of the fields.

The single-factor approach has great merits. It draws our attention to a neglected element, and it simplifies analysis. But when it goes too far, careful consideration reveals glaring logical flaws. For example, White sometimes invokes dubious functional logic: "Northern Europe *had* to develop a new agricultural technique and above all a new plough" (p. 42). Why did it have to? Also, a strict determinist would see the plough as both a necessary and a sufficient cause of agricultural advance, but White seems to shy away from such implications when he states that the plough made highly productive agriculture possible but did not cause it. Finally, in later writings, White (1978) shifted from technological to cultural determinism, invoking Christianity to explain the acceleration of technological progress in the Middle Ages. Can this argument from religion be compatible with his earlier stress on tools?

Let us return to the logic of the plough argument. In effect, White endorses the Whig assumptions that, according to John Staudenmaier (1985 and this volume) pervade technological history. White implies that an autonomous logic of technology pushed European peasants down the path to inexorable "progress" (here, increasing agricultural productivity) We recognize in White's simplified account of the plough the same logic that uncritically invokes "the machine" as the fundamental agent of change in the industrial world.

Discoveries of dynamic agrarian societies that did not need the heavy plough undermine the empirical grounds of White's argument. Italy and Southern France, as David Herlihy (1958) argued, raised production enough to support extensive urbanization after the tenth century A.D., but they used the scratch plough. There, the key was the consolidation of scattered strips of land, not the plough itself.

Still, finding substitute technologies or alternate paths does not completely refute the determinist; it only limits his universality. The determined determinist could admit that only northern European soils needed the plough, if he allowed southern Europe to follow a different path. His logic parallels Alexander Gerschenkron's (1962) thesis about economic development. Strict, single-factor reasoning, embraced by Marx and Walt Rostow, sees English or American private entrepreneurs as the only source of successful industrial capitalism. Gerschenkron argued that not all countries have to follow this route. In Germany or France, banks could substitute for the missing entrepreneurs; farther east, in Russia or China, the state could do the same.

Single-factor theories falter more seriously when critics find that the new technological element already existed in the society. If widespread adoption comes long after invention, then the use of technology requires an appropriate social environment. Therefore, the true determining factor is the social environment, not the technology itself. Hilton and Sawyer (1963) used this tactic in their devastating critique of White, arguing that open-field agriculture existed in Europe for centuries before the invention of the plough. White's other famous thesis, linking medieval knights to the invention of the stirrup, fell to the same critique. White implicitly conceded the point when he tried to explain the lag between the plough's introduction and the completion of the medieval agrarian system. He invoked organizational, not technological, factors: "It may be that the 300-year delay between the arrival of the modern harness and the widespread use of the horse for non-military purposes can be explained by the *practical difficulties* of switching a village from the biennial to the triennial rotation." (p. 73; my emphasis) The vague words "practical difficulties" smuggle in the social relationships which the single-factor thesis drove out.

White himself admits that the plough could succeed only in particular social contexts. Here are several examples. He claims that the new agriculture required high population density, but does not say whether dense settlement followed or grew along with the new plough. If population and technology spread together, neither determined the other. Different methods of surplus extraction by lords, not different technologies, explain the lag of England behind France in the use of the horse: while France moved toward rent collection, England revived the demesne and labor services. English peasants sabotaged their lords' promotion of the horse by refusing to make their plough teams pull any faster than oxen—perhaps the first documented example of peasants' resisting technological innovation with the "weapons of the weak" (Scott 1985). These are strong counterexamples to the irresistible drive of new technology. Still, on White's account (p. 65), peasant resistance could only slow down the technological juggernaut by about a century. White embraces the "railroad train" model of technological advance, in which societies are arranged linearly along the same track, each one lagging more or less behind the next.

Critics deliver the final blow to the single-factor thesis when they unravel the tight linkages on which it is based, pulling apart each of the presumably necessary elements of the medieval agrarian system. They find open fields without heavy ploughs, and peasants who grew spring crops and legumes on two-field systems. Horses did not replace oxen on most fields in England until the sixteenth century, long after the plough came in. Now the advocate of determinism, to save his case, must concede that the system held together tightly only in certain regions. His highly localized determinism deflates his general theoretical claims (Langdon 1986).

Other critics may believe in a single factor but claim that it is not technology. Marx, on Bruce Bimber's (1990) account, was not a technological determinist but a class-struggle determinist. In all debates about agricultural development, three standard alternatives to the technological argument reappear: population (in two variants, Malthusian (1798) and Boserupian (1965)), class structure (in various versions, of which surplus extraction is the simplest), and the market. Not surprisingly, each of these arguments is backed by a great theorist: Malthus, Marx, and Adam Smith, respectively. The application of these theories to agricultural systems varies little over time or place. Langdon (1986, pp. 254–292) summarizes the applicability of these theories to Medieval England. I discuss their applicability to China below. To the extent that they only replace the technological factor by another one, these theories suffer from the same flaws as technological determinism.

Reforming Chinese Agriculture: The Single-Factor Theory in Practice
Single-factor theories often appeal to agrarian reformers. If you focus your efforts on one isolated factor, a hopeless task looks more manageable. Many Americans who visited China in the early twentieth century to end peasant poverty took this approach. In Nanking the renowned John Lossing Buck set up an institute, backed by the Rockefeller Foundation, which surveyed thousands of Chinese farms in order to make them more efficient. Completely ignoring China's political environment, he focused narrowly on farm size and land use. Viewing the Chinese peasant as a mini-capitalist who wanted to maximize income, Buck compared the returns for each farm from hiring labor vs. renting land, growing potatoes vs. growing millet, and so on. He sent students all over the country to conduct surveys and compiled the statistics in a series of large volumes, which are still the most extensive source on Chinese agriculture in the 1930s (Buck 1930, 1937; see also Stross 1986).

Buck aimed to merge research and reform. He thought that by focusing on the narrowest cause of Chinese peasant poverty—how the farmer chose which crop to grow in which field—he could evaluate the efficiency of Chinese farming and make recommendations for improvement. He believed that fields demonstrating the superiority of new seeds and cropping techniques would spread reform by example. In reality, Buck and his students failed. Even though the Rockefeller Foundation invested millions of dollars in peasant surveys and demonstration plots, and the Chinese Nationalist government established commissions on agriculture, desperate poverty persisted.

Why did so little change? Without reliable sources of credit, peasants could not accumulate capital for new production methods. Without security of tenure, they had no guarantees that landlords, warlords, or hungry taxmen would not take away their increased output. The Nationalist government neglected agrarian reform. Instead of suppressing the endemic warfare and banditry that afflicted the countryside in the 1920s and 1930s, it preferred to attack the limited Communist base areas. Finally, the world depression wiped out world markets for China's export crops, leaving farmers penniless.

After 1949, the Communists, having achieved undisputed sovereignty over the mainland, put Buck's correct technical advice into practice. After the Nationalist government in flight from the mainland replaced the Japanese on Taiwan, it too carried out agrarian reform. Like White, Buck uncovered a neglected factor, isolated from its social context, stimulated new thinking, and improved the empirical base. Others, however, applied the new technology under radically different regimes.

Russian Peasants in the Eighteenth and Nineteenth Centuries: An Equilibrium System

Why do peasants resist innovation, and when do they accept it? Michael Confino (1969) investigated these matters in the context of Russia's central agricultural region in the eighteenth and nineteenth centuries. Russian peasants produced food with technology remarkably similar to what western Europeans had used six centuries earlier. They, too, practiced three-field rotation and used heavy ploughs and horse teams, raising mainly rye for the famous Russian black bread. By the eighteenth century, however, this system had lost its dynamism. Advanced western European peasants had escaped from the low yields of the manorial economy into enclosures, individual household farming, cash cropping, and rising productivity. Controversy still rages about the extent, the significance, and even the existence of the "agricultural revolution" of western Europe, but reformers saw Western Europe moving ahead while Russia stagnated (Chambers and Mingay 1966; Morineau 1970). Perceived backwardness impelled Russians and foreigners to look closely at Russian agriculture. Their observations provide the sources for Confino's study.

Why, despite the best efforts of well-intentioned landlords to introduce the latest agronomic knowledge from the West, did Russia's agricultural output fail to grow significantly from the eighteenth century to the middle of the nineteenth? Contemporaries usually blamed the Russian peasant, whom they saw as ignorant, sunk in routine, and bound by tradition. Confino, however, dug deeper. He undermined this stereotype by analyzing the systemic relationships between Russian agricultural technology and its social and environmental context, and he discovered a rigidly determined system of elements held in equilibrium by interacting forces. Contrary to the single-factor model, no one element determined the whole. Even though Confino's study focused on the three-field rotation, he stressed that the field structure did not determine the rest of the system.

Compared with western Europe, the Russian three-field system developed fairly recently. Through the seventeenth century, most peasants either shifted cultivation, burning forests and abandoning them, or practiced rudimentary two-field systems, leaving half the land fallow. Regular three-field rotation began to dominate throughout the central agricultural region only under the more stable political conditions of the eighteenth century. The other elements of the system spread with it; these included population growth, serfdom, the replacement of the scratch plough by the heavy plough and of oxen by horses, and the replacement of scattered hamlets by solidary peasant communities. Economic incentives bound these elements together. Population growth induced peasants to switch to the three-field system for extra productivity; lords bound their peasants to the land and encouraged increasing populations so as to maximize their surplus extraction; peasants gathered in nucleated settlements in order to free large open fields and created village councils to manage cultivation rights.

In addition, distinct features marked off the Russian system from the western European one. Most striking, low-yielding rye almost displaced wheat as the main crop. The much colder climate, the poorer soils, and the lower population density of northern Russia disfavored wheat. Eighteenth-century wheat required more labor than rye. Most important, the three-field system, once it took over, locked in rye and prevented hardier strains of wheat from being introduced:

> . . . Independent of the causes which called for and implanted the three-field system, from the moment it became a regular and regulated system of rotation, it determined not only the framework of rotation, but also the types of cereals cultivated. It was this combined action of the vegetative cycle of different crops, linked to the climate and seasons, with the cultural structure, the result of their combination, which fixed one of the important traits of the three-field system in Russia. (Confino 1969, p. 87)

The field rotation, once in place, thus blocked a more economically rational structure. Industrial designers call such implantation the

"QWERTY phenomenon." The modern typewriter keyboard, with the letters QWERTY on the top row, was designed to prevent the jamming of mechanical keys when frequently used letters were typed too quickly. Even though electronic typewriters have no jamming problem, efforts to replace the QWERTY keyboards with faster ones have not succeeded. "We are committed to [QWERTY], even though it was designed to satisfy constraints that no longer apply. . . . The severe constraints of existing practice prevent change, even where the change would be an improvement." (Norman 1988, pp. 147, 150; cf. David 1985)

In Russia, the costs of getting the entire village commune to agree to a shift of cropping prevented the introduction of new crops with unknown immediate economic advantages. Cultural "locking in" reinforced technological blockage. Rye—celebrated in proverbs as the prime symbol of abundance, the "king of the fields," and the "grandfather of all grains"—had such prestige that few peasants would switch to wheat.

The stabilized system damaged the peasants' welfare. All three-field systems tend to reduce the fodder crop, but in warmer climates animals can compensate for the loss by grazing on the stubble in winter. In snowbound Russia, where animals stayed indoors in the winter, they constantly suffered from inadequate fodder supplies. In the spring, the weakened beasts ploughed poorly and produced little manure, further reducing the productivity of both cereal and fodder fields (Confino 1969, p. 68).

The fusion of these features produced a strongly integrated ensemble: poorly fertilized soils, weak cattle, low output, and rudimentary technology. Labor requirements built in further rigidity. Brief intervals of warm weather demanded short bursts of intense labor, followed by long spells of inactivity in the winter. Peasants concentrated their efforts on rye, the main subsistence crop, at the expense of more productive but subsidiary crops like beans, oats, and buckwheat. Note also that rigidity depended on the absence of extensive market networks; otherwise peasants could have traded for fodder and hired labor from the south during intense labor periods (as they did in China).

Thomas Hughes argues elsewhere in this volume that large interlocking technological structures generate "momentum," which allows them to shape their environment. A look at agrarian systems confirms but modifies his analysis. First, technological brakes are as strong as momentum: large systems stop as well as push change. Second, I must dissent from Hughes's implication that large systems necessarily have greater momentum than small ones. The propulsion or braking of a system of elements depends on the tightness of the links between the elements, not on their complexity. Each primitive Russian field, joining soil, tools, village structures, and customary law, tightly constrained the possibility of change, even on a minuscule scale.

The agricultural system strongly affected social organization, too. The persistence of the redistributive commune (*mir*), of course, struck all observers as a major distinction between Russia and Western Europe. The powerful *mir* reallocated rights to land periodically, so as to ensure the subsistence of all village members. The three-field system did not create the commune, but strongly reinforced it. Scattered strips and periodic redistribution enforced *Flurzwang:* that is, the same crops had to be cultivated on all strips. The *mir* allowed no hedges or walls, hence no individual experimentation or cropping variety. Furthermore, as Lynn White argued, the long, narrow strips went together with the heavy plough.

Confino qualifies his determinism more heavily than does White. He rejects the views of German observers, who attributed these features to *Volksgeist* (we would now say "culture"), but he also rejects Marc Bloch's close connection of the heavy plough with collectivism and the scratch plough with individualism (Confino 1969, p. 113). Confino doubts that the plough itself determined the form of social life; he maintains that it was only one element in an ensemble. Confino's argument defies refutation better than White's. Finding heavy ploughs without redistributive communes, and scratch ploughs with the commune, would not invalidate his thesis that the elements tend to cluster. The light, cheap plough still worked best in light, sandy soils; it allowed peasants greater control of their labor, and it tended to be used for cross-plowing. The heavy, expensive plough, favored by landlords, worked best with teams of oxen and peasants who cleared forests and heavy soils.

Reform efforts, discussed in part 2 of Confino's book, reveal how firmly Russian agrarian society was based on the three-field rotation. Eighteenth-century agronomists criticized peasant agriculture for its low yields (seed yields of only three or four to one, no better than Medieval Europe), its fodder shortages, and its soil exhaustion. They also insisted that the poor productivity of the fields created land scarcity in overpopulated central Russia. Agronomic theory, however, proved difficult to put into practice. Intermingled property rights were the first obstacle. Because the lords' and the peasants' animals were combined in one herd and used a common pasture, lords could not improve the quality of their cattle separately. Similarly, the narrow field strips put lords' and peasants' fields side by side. Landlords tried to expand their own land rights at the expense of the peasants by increasing their own domains, taking over state lands, or driving peasants off the land. Against strong resistance, they tried to convince Russian peasants of the wonders of English clover, tubers, and turnips, and of the English practice of reducing fallow land by feeding animals in enclosed lots. They concluded that none of the new techniques could succeed without the elimination of the three-field system. Partial measures did not work. Almost none of the numerous crop-rotation schemes derived from English and French agronomical theories achieved their goals. Confino found, from 1768 to 1800, only seven cases, limited to single domains, in which the new agronomical systems were actually used. As Buck found in China, demonstrating foreign theories on Russian plots failed to produce agrarian transformation without social and political change. In fact, during this period the three-field system expanded into new areas in western and southern Russia.

Contemporary writers uniformly blamed the peasants for their failure. Reformers attacked the peasant's mentality of blind resistance to change, his simple interest in mere family subsistence, and his ignorance of economic calculations. Confino shows, however, that the main fault lay with the lords, not the peasants. They were the ones who blindly imitated foreign theories, lacking practical experience in agriculture. Peasants did respond to successful demonstrations of new agricultural methods by other peasants, but not to schemes imposed by their exploitative masters. In fact, most of the Russian nobility was far less innovative than the peasantry; unconcerned with agricultural production, they complacently collected their incomes in the old, wasteful ways.

After the failure of the experiments, many of the reformers themselves concluded that the social effects of Russian agriculture made it superior to Western agriculture. At least, they thought, the commune provided for the welfare of everyone, if at a low level. Radical reform, they feared, would produce an uprooted proletariat, a clear source of social discontent. Thus, industrializing western Europe, with its unruly crowds of laboring people, became a negative example for Russia in the nineteenth century.

The failure of ambitious schemes of technology transfer, followed by nativist reaction, is by now a sadly familiar tale. Unlike many failure studies, Confino's account of Russia's abortive agrarian reforms focuses not on the ignorance of foreigners about the society into which they hope to transplant their techniques, but on the ignorance of the native elites about their own society. Wealthy Russian noble reformers had little better understanding than their foreign supporters of the dynamics of peasant society. Even if they ran estates, they did not immerse themselves in peasant life. On the other hand, 97 percent of the nobility differed only slightly from their peasants and had no interest in change. The small elite agitating for change were too distant from peasant life to know how to transform it, while the large mass on the land knew too little about the West's progress to follow it. In both Russia and China, only long and painful immersion in the countryside in the late nineteenth and early twentieth centuries brought the discovery of how to harness peasant forces to revolutionary transformation.

The strong links between field systems, cropping patterns, settlement patterns, and the social arrangements of peasant society indicate that only major social transformation could alter the features of Russian agriculture. All piecemeal efforts at reform ran up against obstacles produced by the rest of the equilibrium system. Creating enclosed fields for crop experiments required breaking up the redistributive commune. This was possible only on newly settled or state lands. The lords opposed the end of the commune as much as the peasants, because they depended on the commune for tax collection and they feared the landless class that the abolition of the commune would produce. Likewise, the three-field system favored low-yield crops, such as rye, which required little labor. Efforts to introduce wheat into the rotation ran up against the scarcity of peasant labor (attacked by lords as "laziness"). Because of the lack of a market for hired labor in agriculture, little could be done, especially since the lords were heavily in debt and lacked the capital for improvements.

The "strength of routine" argument, so often invoked as the reason for the failure of reform (Blum 1961, p. 409; Robinson 1969, p. 244; Confino 1969, p. 323), fails to locate where and how technical change can occur. Comparative study of peasant societies reveals many possibilities. For example, the spread of hired rural labor in China offered farmers the chance to produce more from their fields by relieving labor shortages during busy seasons. Russia's hired-labor market, on the other hand, flourished only outside agriculture. When industrialization began, in the late nineteenth century, peasants did work for cash wages in cities, but until outside forces disrupted the agricultural equilibrium they resisted efforts at change.

Even though it is seldom discussed by historians, technological failure can be just as instructive as success (Staudenmaier 1985, pp. 175–176). Failures of reform efforts indicate both how rigid the technological-social equilibrium is and where its linchpins lie. Reform fails not because change is impossible but because the reformers, bound by their own society's constraints, fail to address the fundamental features. Pragmatically, they attack subsidiary elements instead of taking on the entire system, but in fact they fail to get at the roots of the problem. Conversely, reforms can succeed either because reformers decide to mount a radical attack on the system's deeply rooted problems (the revolutionary path) or because liberal tinkering works best in a loosely linked system. The important link is the elites within the country who try to bring new ideas to their rural countrymen. Ultimate success or failure depends on the connection between the reforming elites and the receiving country's social structure.

According to Confino, the Russian system in the early nineteenth century lacked three fundamental engines of change: an integrated rural marketing system, securely enforced property rights in land, and processes of class differentiation. Increasing population and increasing pressure on land did not by themselves impel change. The lack of rural markets blocked substitute paths of development which could get around climatic and soil limitations. Fodder could not be imported from the south, and hired labor for peak seasonal demand was not available. As long as the commune ensured minimal subsistence to all, no landless class grew to provide a supply of hired labor either for rural proletarian labor or for new factories. Most important, serfdom removed incentives for both lord and peasant to alter the system to their mutual benefit. Given redistribution and lack of rights to his own land, no sensible peasant would invest in risky new crops. Instead he used the "weapons of the weak"—foot-dragging, shirking, sabotage, feigned compliance—to resist new ideas of enlightened masters, just like his medieval English counterpart. All these factors underlay the psychological resistance to innovation by both lord and peasant summed up in the concept of "routine."

The late nineteenth century knocked out the vital props of the equilibrium and set it on a new, unpredictable course. The end of serfdom in 1861—a reform imposed by the Russian state on lords and peasants, against the will of many of them—set the stage for a "psychological revolution" on the land. Now "commune eaters" (*miroedy*) set to work—peasant improvers who formed enclosed plots separate from the commune. As markets began to spread, a hired-labor force grew. Class differentiation did set in—perhaps not as inevitably as Lenin thought, but the process was underway.

Confino's brilliant analysis ranks as one of the most outstanding integrations of technology, society, and culture ever written. He almost achieves the *Annales* project of integration of the totality of a society, incorporating geographical environment, technological base, social structure, and mentality in a unified whole.

For students of technological change, many of Confino's arguments echo those found elsewhere. In this volume, Richard Bulliet argues that "social filters" can prevent widespread diffusion of new technologies. In societies with a large gap between social classes,. innovations developed in one stratum fail to spread because the stratum is despised by other groups or is so insulated from them that its ideas and practices remain unknown. The case of the Russian agrarian reformers confirms Bulliet's insight in reverse. In Russia, the innovative ideas came from a small minority of the agrarian elite, which attempted to spread reforms both to the serfs and to the rest of the nobility. The reformers were not despised, but they were extremely isolated from ordinary Russian rural life. The serfs saw no advantage in pursuing innovations that might only increase their labor burdens; the bulk of the nobility, though living in close contact with the serfs, had very little awareness of developments in the West promoted as the advanced source of new knowledge. High class barriers frustrated the Russian rural reforms.

But let us not assume that only rigidly stratified societies block new technologies. For an American example, consider the recent introduction of information technology into manufacturing (Zuboff 1988, pp. 245–310). The introduction of computer-controlled automation to the paper mill Zuboff studied produced quick improvements in productivity but then reached a plateau. The primary reason for failure to use the information technology to its maximum possible extent was that managers feared losing authority if they allowed operators too much knowledge about the operation of the mill. Social barriers such as the fear of losing their jobs constrain information-technology managers just as much as they did the Russian nobility. New technology always alters power relations; this is the primary source of resistance to it. Fear of change persists, even in progressive America.

The equilibrium analysis demonstrates its superiority to the single-factor analysis in many facets. The interrelationship of all the elements, rather than any single one, determines the whole. The key parameter is the tightness of the linkage between the parts. Some of Confino's claims leave room for debate. Was the Russian three-field system really so rigid (Kingston-Mann 1991)? Why were the same features, much more flexible in western Europe, so resistant to change in the East? The reform efforts of the eighteenth and early nineteenth centuries strongly confirm Confino's thesis of rigid links. Eventually, contemporary Russian analysts concluded that the system could not be reformed piecemeal and went down the revolutionary road.

Twentieth-Century China: Flexible Equilibrium

At first glance, Chinese agriculture appears to contrast with Russian agriculture in every respect. Russia is sparsely populated, China is dense; Russian yields are the lowest in Europe, Chinese yields per acre are near the highest in the world. The Chinese practiced diverse forms of agriculture; they moved early on into cash cropping; there was extensive marketing of produce. Serfdom and other bonded labor, if it ever existed, had disappeared almost entirely by the sixteenth century A.D. Villages had very weak controls over land use. Private-property rights, including written contracts for sales and tenancy, were extensive. And yet, by the early twentieth century, complaints about Chinese agriculture as technologically backward and tradition-bound, and the Chinese peasant's miserable poverty, seem strangely reminiscent of the Russian debate two centuries earlier. Like their Russian predecessors, Chinese intellectuals, some foreign-trained, debated the condition of the Chinese peasantry, surveyed farms, and wrote reformist tracts. Once again, the political agenda of reform often obscured more than it illuminated the real substance of peasant life. Still, in recent years, we have begun to gain a better understanding of the Chinese agricultural system.

Roughly the same explanatory patterns recur here in an Asian context. Single-factor technological determinism is best represented by John Lossing Buck and his agronomy school, as mentioned above. Buck's single-mindedness in focusing on crop selection, land utilization, or household budgets while excluding the social context is vulnerable to similar critiques. First, for centuries agricultural manuals promoted more efficient crop selection. Thirty percent of adult males may have had some basic literacy in the difficult Chinese character script (Rawski 1973); and even if peasants did not read them, others passed on the knowledge. If advanced technical knowledge was available, why was it not adopted? Second, Chinese agriculture is so variegated that almost any conceivable combination of crops, seed, field size, and tools can be found somewhere. It is very difficult to assert inexorable links between elements of the production system in a huge, diverse subcontinent such as China. Russia, arguably, has more uniform climatic and soil conditions, at least in the central region. Buck divided China into eight agricultural regions and evaluated the production methods separately in each region. Although he had his students survey more than 17,000 farms, their regional coverage was quite limited. The surveys covered fewer than five districts in many provinces, and even those chosen tended to be wealthier areas near cities and railroad lines (Stross 1986, p. 184). So the representativeness of Buck's data is questionable, although it is certainly much more abundant than White's medieval data.

Mark Elvin's thesis of the "high-level equilibrium trap" is another variant of the determinism thesis applied to Chinese agriculture. In *The Pattern of the Chinese Past* (1973), Elvin describes the many changes that took place in China from the tenth to the thirteenth century A.D. as "the medieval economic revolution." Along with extensive urbanization, population growth, overseas trade, seagoing ships, and new forms of commercial credit went significant transformation of agriculture. Rice-paddy agriculture developed into intensive production based on careful water management and seed selection. The increased output of the lower Yangtze delta was sufficient to feed both the rural and the growing urban population, probably at a rising standard of living. Technological changes on the land, driven by population increase, were the key to the agricultural revolution. Especially notable among these changes are the widespread use of early-ripening rice seeds introduced from southeast Asia (which allowed double-cropping) and the draining of the swamps of the lower Yangtze delta and their replacement by a finely calibrated water-conservancy system (which supplied just the right amount of water at the right temperature during the growing season) (Ho 1956).

But somewhere between the fourteenth and nineteenth centuries this economic revolution ran out of steam. Elvin rejects common explanations that attribute the slowdown in technological progress to political events, such as the emperor's cancellation of ocean voyages in the early fifteenth century; damaging as this was to overseas trade, it had only a short-term effect. For Elvin, the essence is the culmination and exhaustion of the resources of traditional agricultural technology. By the end of the nineteenth century, Chinese crop production, based on non-mechanized agriculture and non-chemical fertilizers, had reached its limits. Only the modern West could break through this barrier by introducing new methods of scientific agronomy based on the machine and phosphates.

Elvin's explanation shares with Buck's the concentration on technical factors and the denial of the importance of social structure or politics. It is also vulnerable to the same criticisms. If machinery was so vital to improving agricultural yields, why did Chinese peasants not adopt it earlier? Furthermore, the capital requirements of scientific agriculture were not so great, and the gap between traditional and modern techniques not as large as Elvin argues. In any case, China may not have exhausted its traditional techniques: Perkins (1975) argues that increasing yields under traditional technology continued through the early twentieth century (see also Lippit 1987). Even in the recent rural upsurge, rural mechanization, although highly touted by government reformers, has turned out to be only one element, and not the most significant one, in improving agricultural yields. Plastic sheeting, which allows hothouse vegetable growing on the cold north China plains, probably has made more of a difference than tractors in improving peasant livelihood.

Each of the alternative explanations to Buck's and Elvin's technological approach has its adherents. The Malthusian school is best represented by the economist Kang Chao's study *Man and Land in Chinese History* (1986), which takes the extreme view that all the important features of Chinese agriculture can be explained by the ratio of population to land. According to Chao, the Chinese have an ingrained cultural drive, determined by Confucian precepts, to propagate their families up to the limit of subsistence. Whatever gains are made in agricultural productivity per land area will inevitably be negated by demographic increases that keep per-capita output at a constant level. Although wars and social collapse during times of dynastic transition will decimate the population, it will recover eventually, returning to the subsistence level. Chao also, like a good neoclassical economist, posits the existence of an efficient, integrated marketing system all over China which allocates sources of land, labor, and capital to their most efficient uses. Population growth, then, exogenously induced by a cultural imperative, drives the intensification of agriculture, which proceeds to the most efficient possible extent in production per unit area, but inevitably keeps the peasantry at a constant, low per-capita income.

Chao ambitiously encompasses the entire sweep of China's dynastic history, from the second century B.C. to the present, in one elegant model. The simplicity and clarity of the model, however, depart substantially from the facts (Wang 1990; Little 1989). Chao's population estimates depart wildly from the consensus view; he exaggerates the population swings occurring when dynasties collapse because he fails to distinguish between the disappearance of people by starvation, disease, or war and the failure to register for tax purposes. He takes no account of regional variations in agriculture, and he assumes that the logic of the market system floats freely over the landscape, unattached to the local imperatives of soil, water, and temperature. Even more than Buck, Chao pushes a single-factor explanation (in this case a non-technological one) to its extremes.

Chao's explanation is designed to refute explanations that invoke class structure as the explanation for China's poverty. Ever since the 1920s and the 1930s, when Chinese intellectuals simultaneously absorbed Marxist perspectives on historical change and began to seriously investigate the countryside, they have invoked the "feudal landlord system" of classical China as the central cause of China's agrarian stagnation. Surplus extraction forms the core of this explanation: China, it is argued, had highly unequal income distribution, with 30 percent of the net national product controlled by 1.9 percent of the population (Lippit 1987, p. 90). This elite class, primarily urban and comprising merchants and gentry, had no interest in improving agricultural or industrial production, and instead diverted the surplus to luxury consumption. Foreign imperialists, when they defeated the Chinese empire in warfare and moved into treaty ports, only aggravated the process. Their investments mostly extracted wealth from China, it is argued; whatever they did invest in new industry never reached beyond the limited sphere of the modern treaty-port cities (Shanghai, Canton, Tientsin, etc.). Or, in a variant argument, foreign-made textiles flooded rural markets, throwing rural handicraft weavers out of work and increasing their desperation. Only when the Communists took power as a new elite dedicated to using the surplus for productive purposes could Chinese agriculture and industry move forward.

The surplus-extraction school, in its turn, has been attacked most convincingly by neo-Smithians of different stripes, all of whom share an emphasis on the extension of competitive markets as the primary engine of growth (Brandt 1989; Myers 1980; Rawski 1989) For them, China had active, competitive rural markets which satisfied all the basic requirements for the operation of the invisible hand: low barriers to entry, easy availability of information about costs, and no domination of supply or demand by any small privileged group. From the late nineteenth century to the middle of the twentieth, the operation of these free markets produced significant economic growth—as much as 1.5 percent per year from 1870 to 1937, twice the rate of population growth (Brandt 1989, p. 9). The operations of commercialization and specialization in agricultural markets drove economic growth in spite of temporary interruptions by political turmoil, famine, and flood.

Without going into a detailed critique of all these theories, I will simply note that each of them follows the unfolding of a single logic, whether that of technology, that of class, or that of commerce, as the underlying mechanism driving agricultural change. All these authors admit other factors, of course, but essentially only as subsidiary influences. Only very recently have a number of studies sought to develop the alternative represented by Confino: an equilibrium analysis of Chinese agriculture. Of these, Philip Huang's two volumes (1985, 1990) are the most ambitious.

Huang tackles a question very similar to that raised by the Russians in the eighteenth century: Why were the Chinese peasants in the twentieth century so poor? Like the Russians, his implicit mirror for measuring Chinese agriculture is the advanced farming of England and northern France, which seemed to have raised the per-capita output and provided a surplus to fuel urbanization and an industrial revolution. By the 1920s China remained one of the poorest countries in the world. Huang finds the answer in a long-term process he calls "involution" (a term borrowed, but modified, from Geertz's classic 1968 work on Indonesia). Involution in China meant a complex of ecological, technological, cultural, and economic factors that produced, over six centuries, continually increasing yields per acre but constant or declining returns per labor day. Nearly all progress in agricultural output from 1300 to 1980 was due to increased labor input by the growing population, extension of the cultivated area, and increasing yields.

Huang focuses on the failure of "managerial farming" (farming done by a single owner with the use of hired labor). On the North China plain, about 9 percent of the farms were run this way. Single owners or tenant farmers cultivated nearly all the rest. Managerial farms were larger than average. Their owners had more land than they could work themselves, but unlike their Russian counterparts they could take advantage of active rural labor markets. Beginning in the eighteenth century, they developed cash cropping of cotton. Everything seemed to favor China's progress down the capitalist road, just as John Lossing Buck had dreamed. Millions of mini-entrepreneurs could allocate cropping, fertilizer, labor, and capital by the logic of profit, sell their products on the market, and rely on the labor of a burgeoning rural proletariat. Markets, class differentiation, and property rights favored this path. But, as Huang shows, peasants in northern China moved only part way down the road, and peasants in the lower Yangtze delta did not follow it at all. Managerial farms in the north had no major economic advantages over the microscopic plots of owner-cultivators and tenants, for dry farming of wheat and millet provided no great economies of scale. The self-exploitation of family farming negated managerial farmers' slight gains. Owner-cultivator families could always work themselves just a little harder, at an implicit cost below the going wage of hired labor, to compete on the commercial cotton market with potential capitalists.

Hence the technological equilibrium of northern Chinese agriculture lasted for centuries. The ecology of the north China plain—dry fields, fertile soil, tube wells, intense labor, cold or temperate climate—ruled out large economies of scale. A dense, rising population, seen already by the second century A.D., drove the system toward labor intensity, high yields, diverse crop cultures, and moderate commercialization. Further intensification barely provided for new mouths. In addition, because the imperial state based its taxation system on a stable owner-cultivator class, it discouraged the spread of serfdom, plantation labor, and large estates.

Reducing the system to one isolated feature would be misleading. Variations abound. Although China had some large estates, their tenants worked the fields just as the neighboring owner-cultivators did. Enclosures for pasture land, so characteristic of England, never took hold, despite imperial efforts to secure a stable supply of pasture to feed war horses (Smith 1991). Northern Chinese agriculture, a more flexible balance of factors than the Russian system, adapted to gradual economic transformation by introducing markets, population growth, new crops, and hired labor, but preserved its essence.

In the lower Yangtze, a vastly different ecology nevertheless produced the same results. This region featured very high-yielding rice-paddy agriculture (double-cropped with wheat and other winter crops), warm to tropical climates, overabundant water supplies, cheap transport, and extremely high levels of market activity. Here, however, managerial farming, the classic road to capitalism, never developed at all.

To explain south China's failure, Huang invokes a strong cultural constraint: Because powerful biases against women's working outside the home kept them off the hired labor market, they worked at domestic spinning and weaving for zero marginal cost. Textiles became the major marketed crop in the delta. The free labor of Chinese women in their households defeated would-be textile entrepreneurs who wanted to follow the European road of putting-out followed by factory production. No one could profit from cotton if he had to pay male wages on the labor market but sell at prices determined by female domestic labor. Although some weaving began to move into factories in the twentieth century, spinning remained a household enterprise. As in the north, the population grew, markets developed, and new cash crops spread without transforming the mode of production. Per-capita incomes remained low. Huang terms this involutionary process "growth without development."

One limitation of Huang's model is its reliance on an implicit cultural determinism. Huang assumes that strict gender separation compelled women to work at home, preventing them from leaving for factories. It was because they were confined to the home that they worked for low to zero wages and could undermine potential industrial competitors. Huang's assertions are not empirically grounded. In Guangdong, south China, at least, many women did work outside the home. Researchers have only just begun to reveal the multiplicity of gender relationships, which belies facile assumptions of unvarying patriarchy (Gilmartin et al. 1992). They have yet to link their work with agrarian development.

Not even the new Communist state could break the links reproducing rural poverty. The new regime inflicted massive upheavals on rural China from 1950 to 1980, but even collectivization, famine, industrialization, and repeated political mobilization failed to shake the agrarian regime. Women unbound their feet, left home to work in fields and factories, and earned their own incomes. Greater water-control projects, mechanical pumps, and chemical fertilizers brought Buck's scientific agronomy to the fields. Even so, agricultural production grew only as much as the rural labor force. From 1950 to 1980, China's rural population grew from 502 to 761 million, but per-capita income remained constant.

Only since the 1980s has the long-established agrarian system (unexpectedly) collapsed. Huang denies that the victory of free-market family farming accounts for the tripling of rural incomes in the last decade. Instead, collective rural industry, from basket weaving to the processing of hogs and mushrooms, has made the most difference. Finally, Chinese peasants are leaving the farm, working in local factories, earning significant cash incomes, and buying city-manufactured goods. The adventurous ones are striking deals with foreigners from Hong Kong, the United States, Japan, and Austria. Huang believes that China's countryside is finally on the way to capitalist accumulation. This is Huang's most controversial argument. Although most economists predict continually increasing rural incomes for some time, we cannot be sure that a real breakthrough has yet happened. China has had other brief agricultural spurts that have petered out, but this one looks to be longer-lasting.

Huang's interpretation of rural China is far from the last word. Many scholars hotly dispute his conclusions (Perdue 1992; Wong 1992; Gottschang 1992; Myers 1991). I use his work here to demonstrate the advantages of a systemic analysis that combines the constraints of technology, local ecology, and socio-economic factors. Without privileging any one factor, Huang makes a convincing argument for the interconnectedness of the ensemble: a looser, more dynamic ensemble than Russian three-field agriculture, but one that did constrain the development of Chinese agriculture for many centuries.

However, Huang omits other relevant factors. If Chinese agriculture has now broken its historical constraints, clear planning by a centralized state deserves little credit for it. It came, ultimately, from millions of peasants who seized the opportunities finally offered to them when the state lifted Communism's dead hand. But the peasants' rush for unprecedented opportunities built on the achievements of the Communist regime. The campaigns for rural literacy had created a younger generation of peasants, male and female, much better educated than their ancestors—newspaper readers who could take advantage of new commercial opportunities. The ruthless birth-control campaigns of the 1970s inflicted misery on rural women but reduced their total fertility from 6.0 births per woman in 1950 to 2.9 in 1980—a dramatic achievement crucial in preventing demographic growth from eating up productivity gains. Still, until the 1980s, even a literate, birth-controlling, economically conscious peasantry had no incentive to produce above the level equal to subsistence plus a hefty share of state extraction. Separately changing each of the elements in the interconnected system could not alter it fundamentally. Their joint alteration finally broke the mold of traditional agriculture—but in unexpected ways.

The best recent analyses of agrarian societies examine technological elements in a social context. Climate, soil, crops, animals, and ploughs constrain, but do not determine, what farmers grow. Rice-paddy and three-field agriculture fulfill the definitions of technology offered by Hughes and Heilbroner (see their essays in this volume): an interrelated set of elements whose linkage defines the system, and a clear sequence of progressively more advanced ways of extracting resources from the soil. Common elements recur around the world: questions of population growth, soil fertility, class structure, and market development are universal. But the long-term development of different systems does not follow any one simple model impelled by a single logic. As in industrial systems, politics, economics, and culture must be incorporated into the analysis to produce adequate explanations.

The contextual interpretation of technology has much to offer all who examine society, industrial or agricultural. All determinisms limit human freedom, but freedom includes the recognition of necessity. Ironically, the more we discover about human and natural processes, the more potential power we seem to have and the less real control.

Once we recognize that our choices are limited but not totally determined by the natural and social orders we live in, in both industrial and agricultural production systems, we can either surrender to despair or, in a new humility, discover appropriate roles for ourselves in the interrelated structures that humans, natural forces, and their tools create.

Bibliography

Anderson, Perry. *Lineages of the Absolutist State.* Verso, 1979.

Aron, Raymond. *Main Currents in Sociological Thought,* vol. I, tr. R. Howard and H. Weaver. Doubleday, 1968.

Bimber, Bruce. "Karl Marx and the Three Faces of Technological Determinism." *Social Studies of Science* (May 1990): 333–351.

Bloch, Marc. *French Rural Society: An Essay on its Basic Characteristics,* tr. J. Sondheimer. University of California Press, 1966.

Blum, Jerome. *Lord and Peasant in Russia from the Ninth to the Nineteenth Century.* Princeton University Press, 1961.

Boserup, Ester. *The Conditions of Agriculture Growth.* Aldine, 1965.

Brandt, Loren. *Commercialization and Agricultural Development: Central and Eastern China, 1870–1937.* Cambridge University Press, 1989.

Buck, John Lossing. *Chinese Farm Economy: A Study of 2866 Farms in Seventeen Localities and Seven Provinces in China.* University of Chicago Press, 1930.

Buck, John Lossing. *Land Utilization in China* (three volumes). University of Nanking, 1937.

Chambers, J. D., and G. E. Mingay. *The Agricultural Revolution, 1750–1880.* Schocken, 1966.

Chao, Kang. *Man and Land in Chinese History: An Economic Analysis.* Stanford University Press, 1986.

Confino, Michael. *Systémes Agraires et Progrés Agricole: L'Assolement Triennal en Russie aux XVIIIe–XIXe Siecles (Agriarian Systems and Agricultural Progress: The Three-Field Rotation in Russia from the Eighteenth to the Nineteenth Century).* Mouton, 1969.

Cronon, William. *Nature's Metropolis.* Norton, 1991.

Cronon, William. *Changes in the Land: Indians, Colonists, and the Ecology of New England.* Hill and Wang, 1983.

David, Paul. "Clio and the Economics of QWERTY." *American Economic Review* (Papers and Proceedings) 75, no. 2 (1985): 332–337.

Elvin, Mark. *The Pattern of the Chinese Past.* Stanford University Press, 1973.

Geertz, Clifford. *Agricultural Involution: The Processes of Ecological Change in Indonesia.* University of California Press, 1968.

Gerschenkron, Alexander. *Economic Backwardness in Historical Perspective.* Harvard University Press, 1962.

Gilmartin, Christina, et al., organizers. "Engendering China: Women, Culture, and the State." Conference held at Harvard University, Wellesley College, and MIT, 1992.

Gottschang, Thomas. "Incomes and the Chinese Economy: Comments on the Debate." Paper delivered at annual meeting of Association of Asian Studies, Washington, 1992.

Herlihy, David. "The Agrarian Revolution in Southern France and Italy, 801–1150." *Speculum* 33 (1958): 21–41.

Hilton, R. H., and Sawyer, P. H. "Technical Determinism: The Stirrup and the Plough." *Past and Present* 24 (April 1963): 90–101.

Ho, P'ing-ti."Early Ripening Rice in Chinese history."*Economic History Review,* second series, 9 (1956): 200ff.

Huang, Philip C. C. *The Peasant Economy and Social Change in North China.* Stanford University Press, 1985.

Huang, Philip C. C. *The Peasant Family and Rural Development in the Yangzi Delta, 1350–1988.* Stanford University Press, 1990.

Jones, Eric. *The European Miracle: Environments, Economies, and Geopolitics in the History of Europe and Asia.* Cambridge University Press, 1987.

Jones, Eric. *Growth Recurring: Economic Change in World History.* Oxford University Press, 1988.

Kingston-Mann, Esther. "Peasant Communes and Economic Innovation: A Preliminary Inquiry." In *Peasant Economy, Culture, and Politics of European Russia, 1800–1921,* ed. E. Kingston-Mann and T. Mixter. Princeton University Press, 1991.

Langdon, John. *Horses, Oxen, and Technological Innovation: The Use of Draught Animals in English Farming from 1066 to 1500.* Cambridge University Press, 1986.

Lippit, Victor D. *The Economic Development of China.* M. E. Sharpe, 1987.

Little, Daniel. *Understanding Peasant China: Case Studies in the Philosophy of Social Science.* Yale University Press, 1989.

Malthus, T. (1798) *An Essay on Population*. Norton, 1976.

Montesquieu, Michel. *De L'Esprit des Lois*. In *Oeuvres Completes,* volume 2. Gallimard, 1958.

Mokyr, Joel, ed. *The Economics of the Industrial Revolution*. Rowman & Allanheld, 1985.

Morineau, Michel. "Les faux-semblants d'un démarrage economique" ("The False Signs of Economic Development"). *Cahiers des Annales* 30 (1971).

Myers, Ramon. *The Chinese Economy: Past and Present*. Wadsworth, 1980.

Myers, Ramon. "How Did the Modern Chinese Economy Develop? A Review Article" (with reply by Philip Huang). *Journal of Asian Studies* 50, no. 3 (1991): 604–634.

Norman, Donald A. *The Design of Everyday Things*. Doubleday, 1988.

Perdue, Peter C. Review of Philip Huang, *The Peasant Family and Rural Development in the Yangzi Delta, 1350–1988*. *Journal of Economic History* (March 1992): 234–236.

Perkins, Dwight, ed. *China's Modern Economy in Historical Perspective*. Stanford University Press, 1975.

Rawski, Evelyn. *Education and Popular Literacy in Ch'ing China*. University of Michigan Press, 1973.

Rawski, Thomas G. *Economic Growth in Prewar China*. University of California Press, 1989.

Robinson, Geroid T. *Rural Russia Under the Old Regime*. University of California Press, 1969.

Schoppa, Keith. *Xiang Lake: Nine Centuries of Chinese Life*. Yale University Press, 1989.

Schultz, Theodore W. *Transforming Traditional Agriculture*. Yale University Press, 1964.

Scott, James C. *Weapons of the Weak: Everyday Forms of Peasant Resistance*. Yale University Press, 1985.

Smith, Paul. *Taxing Heaven's Storehouse: Horses, Bureaucrats, and the Destruction of the Sichuan Tea Industry 1074–1224*. Harvard University Press, 1991.

Staudenmaier, John. *Technology's Storytellers: Reweaving the Human Fabric*. MIT Press, 1985.

Stross, Randall E. *The Stubborn Earth: American Agriculturalists on Chinese Soil, 1898–1937*. University of California Press, 1986.

Tucker, Robert C., ed. *The Marx-Engels Reader.* Norton, 1972.

Wakeman, F. "All the Rage in China" (review of *River Elegy*). *New York Review of Books,* March 2, 1989.

Wang, Yeh-chien. Review of Kang Chao, *Man and Land in Chinese History: An Economic Analysis. Harvard Journal of Asian Studies* 50, no. 1 (1990): 407–411.

White, Lynn, Jr. *Medieval Religion and Technology: Collected Essays.* University of California Press, 1978.

White, Lynn, Jr. *Medieval Technology and Social Change.* Oxford University Press, 1962.

Wittfogel, Karl. *Oriental Despotism: A Comparative Study of Total Power.* Yale University Press, 1957.

Wong, R. Bin. "Chinese Economic History and Development: A Note on the Myers-Huang Exchange." *Journal of Asian Studies* 51, no. 3 (1992): 600–612.

Worster, Donald. *Rivers of Empire: Water, Aridity, and the Growth of the American West.* Pantheon, 1985.

Worster, Donald, et al. "A Round Table: Environmental History." *Journal of American History* 76 (1990), no. 4: 1087–1147.

Xia Jun, director. *River Elegy (He Shang).* Television film. Transcript translated by U.S. Department of Commerce Joint Publications Research Service, 1988.

Zuboff, Shoshana. *In the Age of the Smart Machine: The Future of Work and Power.* Basic Books, 1988.

Determinism and Pre-Industrial Technology

Richard W. Bulliet

Richard Bulliet takes issue with Robert Heilbroner's "economics-dominated" model of technological change, particularly with the view of pre-capitalist societies as essentially "inconsequential" other than as background for the great drama of industrial capitalism. Bulliet argues, to the contrary, that the study of technological change in pre-capitalist cultures reveals complex scenarios that cannot be explained solely by economic behavior. While political economy had a lot to do with the disappearance of wheeled transportation in the Middle East in the fifth century, Bulliet nonetheless holds that other socio-cultural factors exercise a strong determining influence on technology. He illustrates his thesis by reviewing three instances of technological innovation in the Islamic world (block printing, the harnessing of draft animals, and wheeled transport) that failed to have an immediate transforming social and economic impact. In all three cases, certain "social filters" related to class, race, and lifestyle served to stall the dissemination of "potentially transformative technologies." Rather than being "merely interesting illustrations of the myriad and unsystematizable contingencies of pre-capitalist technology," Bulliet maintains, "they all illustrate some social calculus of technological change that persists alongside economic logic and needs to be taken into account in assessing questions of causality." Bulliet is critical of top-down economic interpretations because he sees them as deflecting analysis away from critical questions of power relationships in society. "Powerful and powerless, Western and non-Western, and male and female, the core lines of cleavage in current historiographic thought," he concludes, "cannot be swept aside by the iron hand of economics."

In reassessing his 1967 article "Do Machines Make History?" Robert Heilbroner observes: "If we wish to study a society unfamiliar to us, the best place to start is by grasping its material life. To understand the historical significance of Eileen Power's peasant Bodo, of Mantoux's Arkwright, or of Marx's Moneybags we must first become acquainted with the material circumstances of their lives." But he then goes on to argue that "pre-capitalist technological impingements do not affect their societies with the 'logic' that only comes with capitalism's translation of use values into exchange values. The impact of pre-capitalist technical change therefore appears more contingent, less open to systematic elucidation, than when an economic force field guides its applications and consequences. We can say many more things about the 'path' of technical change in the United States in the nineteenth century than about its course in ancient China or the Roman empire. Perhaps there are other logics that would enable us to describe the interaction of technical change and social consequence with a generality equal to that of economics, but we do not know of them."

As one of only two contributors to this book whose work focuses on pre-capitalist or non-Western technological history, I am unsettled at finding the technical environment of pre-capitalist humankind first granted primacy as a necessary precondition for historical understanding and then dismissed as contingent and unsusceptible of systematic discussion—as a set of stage props necessary to the setting of a drama.

In 1969 I wrote an article that attributed momentous technological change in the pre-capitalist period to a "logic" of economics.[1] I argued that a 20% cost differential between transportation by oxcart and transportation by camelback, as testified to by the Roman emperor Diocletian's edict on prices (ca. 300 A.D.), combined with aggressive fiscal measures taken by camel caravaneers to suppress competition from carters, was the major factor in the disappearance of wheeled transportation throughout the Middle East. This episode in technological history meets certain criteria laid down by Heilbroner. The socioeconomic transformation was large in scale, covering a swath of

1. "Le chameau et la roue au Moyen-Orient," *Annales: Économies, Sociétés, Civilisations* (1969): 1092–1103; English version publised in *Social Historians in Contemporary France: Essays from Annales* (Harper & Row, 1972); argument expanded in *The Camel and the Wheel* (Harvard University Press, 1975).

territory far larger than Europe. The absence of wheeled transportation, which lasted from (roughly) the fifth to the nineteenth century, exerted a powerful force upon events. It altered relations between pastoralists and settled people and between Arabs and non-Arabs. It conduced toward the development of urban circulation patterns that favored tightly packed building, narrow irregular streets, and a pedestrian scale and ambience. It militated against the construction of roads; hence, by the dawn of modern times, it deprived the region of an important infrastructural element. And it inhibited the maturing of vehicle-related crafts, which in Europe progressively led to the development of carriages, trains, bicycles, automobiles, and aircraft.

By this hypothesis, therefore, economics operated as a "maximizing" force within a substantial segment of the medieval Middle Eastern economy—an economy that has also been described as resting on merchant (though not industrial) capitalism.[2] Yet at a subsequent stage, the consequences of this technological transformation, which had once been economically advantageous, severely retarded the economic development of Middle Eastern society.

If this description of the history of Middle Eastern transportation technology is at least roughly sound, several questions must be asked of Heilbroner's formulation. Is pre-capitalist technological history indeed generically different from capitalist theory? If it is different, do its contingency and its lack of system render it inconsequential except as historical background and for its occasional, presumably unpredictable, "catalytic" effect? On the other hand, if it is not so much different as more complex, it is possible that forays into pre-capitalist technological history can help refine the economics-dominated model Heilbroner proposes for modern times? Finally, is it possible that "How do machines make history? Through the iron hand of capitalist economics" is a historical construct of limited applicability that serves as much to obscure the nature of technical change as to illuminate it?

These questions cannot all be dealt with here. I shall focus, through three examples, on a single aspect of pre-modern, non-Western technological history. Having summarized the hypothesis that, in the case of the competition between wheeled vehicles and burden-laden cam-

2. Maxime Rodinson, *Islam and Capitalism* (University of Texas Press, 1978); Mahmood Ibrahim, *Merchant Capital and Islam* (University of Texas Press, 1990).

els, economic "logic" (pre-capitalist though it may have been) powered a long-lasting and transformative change in the economy and the technological environment of the Middle East, I shall concentrate on counterexamples—episodes in technological history that could have had, but did not have, transformative social and economic impact. Though it might be argued that this will only confirm Heilbroner's assertion that pre-capitalist technological history is too contingent for systematic study, the conclusion I shall eventually argue is that the "few well-defined behavioral vectors" Heilbroner seeks in his quest for a systematic approach to technological change may not all be economic, in either pre-capitalist or capitalist times. Alternatives may, as Heilbroner says, seem "impossible to imagine," but this should not inhibit us from trying when we find the capitalist model too restrictive or unilluminating.

Technological boundaries of one sort or another exist in all periods from paleolithic times onward. The dominance of a particularly beneficial technology in one context and its absence in another is often attributable to environmental constraints or to lack of contact, but in other instances it is difficult to account for a technological boundary. The Silk Road, for example, served as a diffusion route for innumerable technologies between the third century B.C. and the fifteenth century A.D. Paper making, printing, silk production, and a long list of agricultural products found their way westward from China, but many other technologies did not. Middle Eastern travelers to China brought back citrus seeds but not soybeans. They watched the Chinese efficiently transport all manner of goods by wheelbarrow but then went home to a society that transported everything on the backs of animals or humans.

To take an example of societies in even closer contact: Muslims in northern India made efficient use of camel carts, which are still an important part of the transportation economy of Pakistan and India. They were also in constant contact and cultural exchange with Muslims in Afghanistan, Turkestan, and Iran. But camel carts never came into common use west of the Indus valley, at least until the twentieth century, despite the plenitude of camels throughout the Middle East.

The essentialist idea that technological boundaries coincide with cultural boundaries—roughly what Heilbroner implies by the word "routine"—is too simple and circular to account for the vagaries of technological diffusion. Some things move from one culture to another; others do not; but the cultures cannot be defined by what crosses boundaries and what does not. The soybean was no more intimately bound to Chinese culture than the silkworm. On the other hand, theories of economic utility similarly lack predictive power, since less efficient technologies can rub shoulders with more efficient technologies for centuries without the differential in efficiency causing the former to give way to the latter. There is no resource- or terrain-based explanation for the absence of camel carts west of Pakistan. More sensitive considerations of causality are needed to counteract the tendency toward cultural determinism, on the one hand, and technological determinism, on the other.

As a contribution toward broadening our conception of causality, hopefully with implications beyond the pre-capitalist era, this essay will focus on technological disjunctures within what are perceived to be coherent cultures. I will use three case studies to support the argument that social groupings by class, race, educational background, etc. act as social filters for technological change, determining to a substantial degree what techniques disseminate and how rapidly they do so.

The cultural preferences of one social group may fail to find favor, or may find favor only selectively, among other groups, not because of the intrinsic qualities of those preferences, but because the groups' social boundaries allow little or much to pass through, depending upon which group is on the other side of the boundary. Moreover, as groups change status in relation to one another, the filter between them may become more or less permeable.

The history of printing in the Middle East provides our first example. Specimens preserved in museums and libraries and found in datable archaeological excavations prove that block printing existed in the Middle East as early as the ninth or the tenth century and remained in use until the fourteenth century, when it apparently died out.[3] Yet it is clear that, despite a longevity of four or five centuries, this early printing technology had little or no impact on the general culture of medieval Islam. Indeed, it had so little impact that its existence, in the form of extant specimens, was rediscovered

3. A full discussion of this history, with additional bibliography, is contained in Richard W. Bulliet, "Medieval Arabic *Tarsh:* A Forgotten Chapter in the History of Printing," *Journal of the American Oriental Society* 107, no. 3 (1987): 427–438.

only a century ago, and descriptive references to it in medieval texts were not recognized until very recently. The significance of this episode in the history of printing stands out when comparison is made with medieval Chinese, Korean, and Japanese societies, which were truly transformed by the invention and dissemination of block printing during the same period.

Two lines of Arabic poetry provide the evidence for the working of a social filter in this peculiar history. Abu Dulaf al-Khazraji, an Iranian poet of the tenth century, wrote "Among us, without publicity or boasting, is the engraver of printblocks [*tarsh*]" and Safi ad-Din al-Hilli, a Syrian poet of the fourteenth century, wrote "How many times has my hand written, by printblock [*tarsh*] of tin, Syriac followed by the language of phylacteries." These are the only appearances of the word *tarsh* so far discovered, and it is apparent that it was a rare word in poetic circles because al-Khazraji explains its meaning in a long gloss. Significantly, both poems are devoted to describing the activities and jargon of the beggars, confidence men, and itinerant entertainers that made up the underworld of medieval Muslim society, the subculture known as the *Banu Sasan.* Abu Dulaf says: "The engraver of *tarsh* is he who engraves molds for amulets. People who are illiterate and cannot write buy them from him. The seller keeps back the design which is on it so that he exhausts his supply of amulets on the common people and makes them believe that he wrote them. The mold is called the *tarsh.*" Since almost all surviving specimens of medieval Muslim printing are indeed amulets—that is to say, long, narrow (2 × 17 inches in the case of a specimen in the Columbia University library) strips of paper meant to be rolled, inserted into a cylindrical metal amulet case, and hung around the neck—their identity with what Abu Dulaf describes seems certain.

The technology for producing amulets in this manner became quite sophisticated. While some specimens are evidently printed from woodblocks, others appear to be printed from tin plates, as al-Hilli's reference to printblocks of tin implies. The black-on-white printing on one specimen bears 11 or 12 lines of writing per inch, with the individual letter size consistently 1 to 2 hundredths of an inch. Not only would such minute writing have taxed any woodcutter to the extreme, but the small, perfect circles that occasionally punctuate it would have been almost impossible, since the block cutter would have had to chisel away the surrounding wood to leave the circle in relief. More likely, the block maker incised his message on a tablet of moist clay, just as his technological forebears had done in ancient times with cuneiform writing, and then made a metal plate from the dried tablet by pouring molten tin on it. This reversed the writing on the plate so that it came out properly once the plate was inked and impressed on paper.

Though the details of this technology need further clarification, the surviving specimens, along with the verses cited, testify to its longevity, its popularity within a certain social group, and the technological skills of the printers. Yet the entire technology was so encapsulated within the subculture of the underworld that learned members of the higher social orders did not even know the word for print block. Moreover, even though religious texts were among the earliest items printed elsewhere in Asia, medieval Muslim printing seems to be an indigenous invention, parallel to and simultaneous with, but not directly dependent upon, what was being developed in China. This is apparent not only from the use of metal plates in the Middle East but also from the fact that other Chinese technologies and styles, such as papermaking and certain pottery glazes, were well received by the dominant strata of Muslim society. It is difficult to see how one important Chinese technology could have become popular among beggars and scoundrels without the upper class, which avidly welcomed the related technology of papermaking, even being aware of it.

By the time letterpress printing was introduced to the Islamic world, in the eighteenth century, all memory of earlier Muslim block printing had disappeared.[4] Strong religious opposition curtailed its spread and restricted it to non-religious texts. That the new technology was an import from Europe, which was then seen as an enemy, certainly acted as a strong social filter. The vested interests of scribes also played a role; however, purely religious opposition may not have been inevitable, since one of the medieval block-printed texts appears to be a page from a Quran, though it is printed on only one side and thus was not part of a bound book.

The primary reason for the limited scope of medieval Muslim block printing, therefore, seems to be that it was developed within an underworld subculture that was socially cut off from the mainstream

4. G. Káldy-Nagy, "Beginnings of the Arabic-Letter Printing in Muslim World," in *The Middle East: Studies in Honour of Julius Germanus*, ed. G. Káldy-Nagy (Budapest, 1974).

of literate society. Even if someone in the subculture had recognized the broader potential of printing, the social filter would have prevented him from exploiting the idea in his own society. The possibility of exploitation in another society provides a tempting explanation for the origin of block printing in Europe in the fourteenth century. Playing cards, called *tarocco* in Italian, are among the earliest known European block prints, and hand-painted playing cards are attested in the Muslim world as early as the twelfth century. It is possible, therefore, that the Arabic *tarsh* entered Europe in this form along with the social flotsam of the late Crusades, and that the word *tarocco* betrays this origin. Unaffected by the Muslims' culturally constructed distaste for the beggars and con men of the *Banu Sasan,* the Europeans may have recognized the potential of printing and felt culturally free to develop it.

A second example of a technology constrained by an impermeable social filter comes from the history of transportation and agriculture in North Africa.[5] From the earliest indications of wheeled transport in Europe and the Middle East down to imperial Roman times, the invariable approach to harnessing animals to carts and wagons was in pairs. The origins of this technology probably lay in the yoking of oxen, since their lowered heads and protruberant thoracic vertebrae make yokes an efficient harnessing device. The yoke adapts poorly to equids, however, since they carry their heads more upright and do not have a bony hump over their shoulders. A strap around the neck and a second one behind the forelegs suffice to hold the yoke in place, but they transfer the point of traction to the animal's throat. Potential strangulation thus serves to limit the weight of the load to less than the animal's full tractive power.

When wheeled transport was introduced into China across Central Asia, the yoke technology for bovids and equids accompanied it; but the Chinese devised new methods of harnessing. One method was to put a yoke across the shoulders of a single bovid and attach it to a plow or a cart by traces. Another was to run a shaft or trace along each side of a single horse and attach them to a breaststrap or horse

5. A full discussion of this history, with additional bibliography, is contained in *The Camel and the Wheel,* particularly chapters 5 and 7, and in Richard W. Bulliet, "Botr et Beranès: hypothèses sur l'histoire des berbères," *Annales: Économies, Sociétés, Civilisations* (1981): 104–116.

collar. The equid methods lowered the point of traction and allowed the horse to use its full force without strangulation.

Lynn White, in *Medieval Technology and Social Change,*[6] highlighted the importance of breaststrap and horse-collar harnessing in the increase of agricultural productivity in late-medieval Europe. White and others asked where and how the new technology originated; they found plausible, if not certain, connections with China by way of the peoples of Central Asia. This left unanswered, however, the question of why the preceding societies of Europe and the Middle East had never hit upon so simple a concept as efficient single-animal harnessing.

Pictorial evidence from Tunisia and Libya, unknown to White, now confirms that efficient single-animal harnessing existed in Roman North Africa. Several bas-reliefs show single camels harnessed for plowing using a lowered point of traction and the equivalent of a breaststrap, and Roman lamps from Tunisia preserve clear pictures of a single horse pulling a cart by means of a horse-collar harness.[7] Moreover, the same harnessing technology for camel plows, camel carts, and horse carts is still in use in rural parts of Tunisia, apparently a technological survival from Roman times.

Why a highly productive part of the Roman world, in close proximity to Rome itself, made use of efficient draft harnessing without its having a broader impact on Roman transportation technology is a question that takes on weight when compared with the profound impact of the same technology, deriving from different sources, described by Lynn White. One answer is that the Tunisian and Libyan practices may, in fact, have had a broader impact in southern Europe, where the history of harnessing has not been as thoroughly studied as in the north. But even if improved cart harnessing did spread somewhat, improved plow harnessing certainly did not.

The social filter formed by the ethnic boundary between Berbers and Romans seems to have been crucial in maintaining the encapsulation of efficient plow harnessing in North Africa. Camels began

6. Lynn White, Jr., *Medieval Technology and Social Change* (Oxford University Press, 1962).

7. Besides the lamp depicted in figure 98 of *The Camel and the Wheel,* two more lamps showing carts drawn by single horses using efficient collar harnesses may be found in Jean Deneauve, "Note sur quelques lamps africaines du IIIe siècle," *Antiquités Africaines* 22 (1986): 152 (figures 12 and 13).

to become commonly available from desert tribes in the first century A.D. Not a native North African animal, the Arabian camel had reached the peoples of the southern Sahara via the Nile valley two or three centuries earlier. There the animal complemented the cattle and horses of earlier pastoral peoples, and its capacity to endure Saharan extremes of heat and aridity fostered an expansion of Saharan nomadism.

The Saharan tribes that bred and used camels were ethnically and linguistically related to the Berbers who worked on Roman farms in southern Tunisia and in Libyan Tripolitania. Without more concrete historical data, the process by which the sedentary, semi-Romanized, agricultural Berbers adopted the strong new animal of their desert kin cannot be detailed. What is evident from the technology, however, is that the Berbers devised their own techniques of saddling, harnessing, and overall animal utilization. The technology attested in historical sources and at the present time in the southern Sahara is entirely different from that in the north.

Berber farmers or farm workers, therefore, seem to have invented the technique of harnessing a single camel to a plow. They are also probably responsible for inventing the camel-cart harness still in use in Tunisia, since the technical design and vocabulary are the same for both. It seems likely, therefore, that the efficient horse-cart harness shown on the Roman lamps was their invention too.

Once invented, however, efficient harnessing did not spread, at least not for plowing. As Roman agricultural enterprise in North Africa declined, some semi-Romanized Berbers developed their own pastoral, nomadic society based upon camel herding. Entirely different in animal technology from the Tuaregs and other peoples of the southern Sahara, with whom they otherwise shared ethnic and linguistic kinship, the northern Berber tribes, known generally to the seventh-century Arab invaders as the *Butr,* spread along the Saharan fringe of Algeria and Morocco and slowly achieved political dominance.

To judge from twentieth-century ethnographic evidence, wherever the Butr tribes went they took with them the practice of plowing with a single camel. Despite their devotion to nomadism, they would plow land and sow a crop in likely spots near various Saharan oases and return months later to harvest the grain. Yet the oasis-dwelling agriculturists—a population of uncertain origin but presumably originating from the southern Sahara, or even further south—never adopted the plow. For centuries they have preserved the hoe technology of the southern peoples—the Tuaregs do not even have a word for "plow"—despite the regular presence of plow-using nomads on the fringes of their settlements.

The Romans' disinclination to adopt the improved technology of an alien North African population that was socially and politically inferior may well have been due to more than the social filter represented by the ethnic boundary, as may the distaste of later Saharan oasis dwellers for the plows of the Berber nomads. But it is difficult to discount the ethnic social filter as a factor in the encapsulation of these useful technologies, particularly when separate bodies of evidence from two widely separated periods point to the same result.

A third example of socially encapsulated technology has to do with camels more generally.[8] The one-humped camel, also known as the dromedary, is native to the Arabian peninsula. Its distinctive adaptations to the torrid desert habitat left it virtually without enemies in this environment. The resulting lack of fear probably contributed to its domestication, which, on currently available evidence, probably took place in the fourth millennium B.C. in southern and eastern Arabia. Despite this rather late domestication, no wild dromedaries exist today, nor have any been reported for at least the past fourteen centuries. The reason for this is that there is probably no major psychological difference between the tractable but independent domestic camel and its fearless wild ancestor.

The range of the two-humped (Bactrian) camel originally encompassed Iran and Central Asia, though skittish and elusive wild specimens survive only in the Mongolian desert. Compared with the one-humped camel, the wild two-humped camel seems always to have been quite rare. Its strength was recognized early on, however, and domesticated animals were used for pulling carts in Central Asia as early as the third millennium B.C. Thus, it was probably first domesticated around the same time as its one-humped cousin.

The early harnessing of the two-humped camel for draft purposes reveals a desire in ancient Central Asian society to utilize the animal's great strength and endurance, and camel carts continued to use in that region into the twentieth century. Yet a succession of neighboring Middle Eastern societies from ancient times down to the thirteenth century A.D. chose not to avail themselves of this useful technology.

8. For a full discussion see *The Camel and the Wheel,* chapters 1–4, 7, and 8.

In fact, camels seem to have been almost completely absent from the Middle Eastern economy, except as desert pack and riding animals, until the Roman period, after which they became almost ubiquitous in settled lands both as transport animals and as a source of power for wells, mills, and irrigation devices.

What prevented camels from being introduced into the general economy of the Middle East in ancient times was the social filter represented by a perceived dividing line between barbarians and civilized people. Only when the Arabs ceased being viewed as barbarians did the animals they bred become generally acceptable to other social groups. Camel-breeding Arabs first appear in history in bas-reliefs of battle scenes from the Assyrian Empire. Not surprisingly, they are always depicted losing to the Assyrian warriors. Of greater interest in the scenes is the poverty of Arab material culture. The Arabs fight with flimsy bows and arrows rather than with swords or spears, apparently because of a lack of metal technology. They wear simple short skirts, and their heads and feet are bare. And aside from the odd cooking pot, their tent furnishings appear negligible. In other words, their material culture is markedly simpler and poorer than that of the Arabs depicted in the immediately pre-Islamic period, who have flowing garments, metal weapons, and war horses.

The primitive level of Arab material life in Assyrian times corresponds to a general fear of the desert in the ancient Middle East. As desert dwellers, the Arabs were not a welcome part of ancient Middle Eastern society or an integral part of the economy. Even camel caravans seem to have been little known or used prior to the Hellenistic period.

What seems to have changed the status of the Arabs was the rather sudden access to wealth brought about by an enhanced ability to control desert trade. Sometime around the second century B.C., a new camel saddle design became prevalent in northern Arabia. Unlike its predecessor, which continues in use to this day in southern Arabia, the North Arabian saddle situated the rider on a firm wooden framework on top of the animal's hump. By contrast, the South Arabian saddle seats the rider behind the hump on a much more precarious perch. Though the details of historical interconnection are lost, the fact that the new saddle enabled the rider to slash or stab down from a commanding height seems to have greatly improved the Arabs' fighting capacity. At roughly the same time that the new saddle was introduced, bows and arrows gave way to swords and spears, and the advantage of great height is specifically mentioned in the sources.

The rise of Petra in southern Jordan, the first Arab caravan city, was due at least partly to the newfound military capacity of the Nabataean Arabs, and Petra rapidly became noted for its wealth and luxury, as did Palmyra, in the Syrian desert, somewhat later. The material culture of the northern Arabs improved markedly, as the ruins of Petra and Palmyra amply testify. With wealth came social acceptance. Arab merchants and caravaneers became part of urban society in the cities bordering the Syrian and Arabian desert, even attaining, in the person of Philip the Arabian, the ultimate status of Roman emperor.

The factors that led to the pack camel's subsequently supplanting the oxcart were partly rational (e.g., economic efficiency) and partly political, the political aspect deriving from the desire of the wealthy Arab merchants to monopolize the carrying trade. In the context of the argument being presented here, however, what is notable is that these "rational" factors—cheap desert grazing, shortage and hence high cost of wood for carts—could as easily have come into play a thousand years earlier. Camels had long been abundant in the desert. The technology for harnessing their energy had been available and cheap. And there is nothing to suggest that the impoverished Arab tribesmen would have been reluctant to sell some of their livestock.

The significant change that opened up the transportation and labor economy of the Middle East to camel utilization was the rise in the social status of the Arabs and the consequent erasure of the social barrier separating the people of the desert from the settled populations. In this case, it was not ethnicity but lifestyle and the unwillingness of settled people to form commercial networks with impoverished nomads that served as a social filter for limiting the dissemination of a technology that, once adopted, changed Middle Eastern society profoundly by turning it from a wheeled society into a pedestrian society.

Each of the three examples sketched above has dealt with a potentially transformative technology. Printing, efficient draft harnessing, and wheeled transport have all been portrayed as crucial elements in describing one or another historical society. Numerous books have been written exploring the origins of all three technologies, on the assumption that pinpointing origins is the historian's most important task.[9] The subsequent dissemination of such valuable technologies has almost been taken for granted.

The conclusion to be drawn from the examples given here, however, is that origins tell only part of the story. To be sure, many people who have had an idea that went nowhere in their own lifetimes have been described as being "ahead of their time" or being frustrated because conditions were not "ripe." No doubt many potentially useful technologies have gone unrecognized at different periods of history because of microhistorical circumstances, such as personal relations or local economic conditions. In the cases under examination, however, the technologies were developed and used efficiently for long periods in close proximity to people who could have adopted them if they had chosen to. What prevented them from doing so were differences in class, ethnicity, and lifestyle.

To return to the more general issues with which I began: The question is whether the examples presented here are merely interesting illustrations of the myriad and unsystematizable contingencies of precapitalist technology, or whether they all illustrate some social calculus of technological change that persists alongside economic logic and that needs to be taken into account in assessing questions of causality. The simplest formulation of such a calculus might be that technological innovations diffuse downward, in terms of relative social status, but not upward. At first glance, this might seem merely to identify a condition of diffusion rather than a motor. But if one were to pursue Heilbroner's effort to explore the history of technology from the standpoint of "modern" historiography, one would have to go beyond the *Annales* approach that he voices to the more recent historical conceptualizations associated with Antonio Gramsci, Michel Foucault, and the "subaltern" historians. This approach to history focuses on power relationships, and upon control of discourse as an implement of power and as a means of distinguishing the powerful from the powerless. Though Marxist in origin, it escapes the narrow limitations of class analysis and concentrates on the formation and operation of hegemonic discourses—among them the idea of capitalist economics

9. See White, *Medieval Technology and Social Change;* Stuart Piggott, *The Earliest Wheeled Transport: From the Atlantic Coast to the Caspian Sea* (Cornell University Press, 1983); Thomas Carter, *The Invention of Printing in China and Its Spread Westward,* second edition (Ronald Press, 1955).

as (to use Heilbroner's words) the only "force field . . . emanating from the technological background to impose order on human behavior in a manner analogous to that by which a magnet orders the behavior of particles sprinkled on a sheet of paper held above it." From this point of view it might well be argued, and exemplified from the technological history of the Third World in the twentieth century, that an insistence on economics as the only imaginable mechanism for explaining technological change in the capitalist era is as much an effort to deflect an analysis of power relationships in technological history as it is an effort to answer the question of technological determinism. Powerful and powerless, Western and non-Western, and male and female, the core lines of cleavage in current historiographical thought, cannot be swept aside by the iron hand of economics. Yet it may require further explorations of pre-capitalist technological history to show the way to a more complete explanation of technological change in modern times.

The Political and Feminist Dimensions of Technological Determinism

Rosalind Williams

Like most of the contributors to this volume, Rosalind Williams is critical of scholars who emphasize "economic agendas and rational motives" when discussing processes of technological change. The source of such thinking is to be found in the Enlightenment conception of progress, which, she maintains, fostered the belief in technological determinism. The argument that technology is inherently rational disturbs Williams because it obscures the fact that technological systems can be, and often are, designed for authoritarian purposes of control and domination. Drawing primarily on the writings of Lewis Mumford, she maintains that we need "to understand the motives behind the construction of powerfully determinative technologies." She also points to the fundamental "dissonance between technological determinism and a feminist understanding of history," noting that, while women have played "highly significant" roles in the development of "democratic biotechnics," they have been "routinely excluded from the creation and operation of authoritarian monotechnics." A feminist perspective, by contrast, recognizes the inextricable relationship that exists between the social and the organic, between human and nonhuman nature. According to Williams, the recognition of such interdependencies is essential. "We dwell in an environment where natural and technological processes have merged," she observes. Any theory of history that assumes the contrary "is plainly unrealistic and simplistic."

The many varieties of technological determinism can be reduced to a three-word logical proposition: "Technology determines history." In four of the best-known overviews of the subject, Donald MacKenzie, Langdon Winner, Robert Heilbroner, and Bruce Bimber all get around to analyzing each of these three words.[1] Let us begin with this approach and see how far it takes us.

What is "technology," for starters? There seems to be general agreement that any definition of technology must begin with material objects, but in many cases the definition extends well beyond that material core. Because the Marxist term "forces of production" includes labor power, MacKenzie notes, it "admits conscious human agency as a determinant of history; it is people, as much as or more than the machine, that make history."[2] Cultural critics such as Jacques Ellul and Lewis Mumford have argued that knowledge and ideology are inherently part of the meaning of "technology."[3] In recent years, the meaning of "technology" has been broadened as historians have come to favor the "technological system" rather than the "machine" or the "invention" as the basic unit of analysis.[4] The concept of the technological system has tended to expand imperialistically into social, cultural, economic, and political domains. Even the nonhuman natural environment may be incorporated into the technological system as a resource, as a sociologized actor (which "cooperates" or "resists"), or as a dump site.[5]

1. Donald MacKenzie, "Marx and the Machine," *Technology and Culture* 25 (July 1984): 473–502; Robert Heilbroner, "Do Machines Make History?" *Technology and Culture* 8 (July 1967): 333–345 (also in this volume); Langdon Winner, *Autonomous Technology: Technics-out-of-Control as a Theme in Political Thought* (MIT Press, 1977); Bruce Bimber, "Karl Marx and the Three Faces of Technological Determinism," *Social Studies of Science* 20 (May 1990): 333–351.

2. MacKenzie, p. 477.

3. See Kenneth Keniston, "Defining 'Technology'," *STS News*, Massachusetts Institute of Technology, (March 1990): 1–3.

4. See the discussion of the three levels of technology (objects, processes, and knowledge) on pp. 3 and 4 of *The Social Construction of Technological Systems*, ed. W. E. Bijker et al. (MIT Press, 1987).

5. On the socialization of nature see Rosalind Williams, "Cultural Origins and Environmental Implications of Large Technological Systems," *Science in Context* 6, no. 2 (1993): 75–101.

For example, the concept of a "Fordist system," or "Fordism," includes much of what used to be called "social context"—management strategies and modes of labor control, as well as techniques of mass consumption.[6] But the core of this multifaceted system is still commonly assumed to be the assembly line. The defining center of the conceptual constellation is technical; it is the hard fact to which the other social, cultural, and natural "factors" are "related." Rather than being said to "act upon" society, the technological system simply incorporates it. By encouraging us to think in terms of a material core dominating nonhuman or immaterial "other factors," the language of technological systems can become a covert form of technological determinism.

Another source of confusion about the term "technology" is the common assumption that certain technologies are more primary, more significant—more determinative, if you will—than others. For example, steam engines and automobiles are tacitly considered more powerful as determinants than household or entertainment technologies. The division of production into primary, secondary, and perhaps tertiary sectors can be especially befuddling when applied to a late capitalist economy, where the proportion of primary production in the traditional sense of sustaining life is so small. As Raymond Williams observes, "By the time you have got to the point when an EMI factory producing discs is industrial production, whereas somebody elsewhere writing music or making an instrument is at most on the outskirts of production, the whole question of the classification of activities has become very difficult."[7]

In the same discussion, Williams takes note of similar ambiguities surrounding the crucial verb "determines." In Williams's view, the common assumption that determination is equivalent to limitation reflects a bourgeois view of society as a system of constraints on a supposedly free preexisting individual. Williams insisted that determination is also pressure, in the sense of ongoing processes that may be internalized far below the level of consciousness.[8] Although Wil-

6. See, for example, the chapters on Fordism in David Harvey's book *The Condition of Postmodernity* (Blackwell, 1989). Harvey, however, does not succumb to the covert technological determinism described here.

7. Raymond Williams, *Politics and Letters: Interviews with the New Left Review* (Verso, 1979), p. 353.

8. Ibid., p. 356. One of the best brief summaries of the meaning of deter-

liams does not invoke Antonio Gramsci in this discussion, elsewhere he reminds us of the importance of Gramsci's analysis of hegemony for any thorough understanding of determinism. Not only is the range of choices extremely limited in any historical situation, Gramsci reminds us; that situation also powerfully shapes the minds of those who choose. "[Hegemony] is a set of meanings and values which as they are experienced as practices appear reciprocally confirming. It thus constitutes a sense of reality for most people in the society. . . ." The most fundamental understanding of what is possible and what is not—of what is determined and what is not—is bound up with the hegemonic order.[9]

A more traditional way of distinguishing various levels of determinism is to speak of "hard" and "soft" varieties. Such distinctions are inherently vague, and they persist when other verbs ("influence," "shape," or, more grotesquely, "impact") are substituted for "determine." The problem of establishing levels of determination is further complicated by the fact that in science—from which concepts of determinism are so largely borrowed, and which is also divided into "hard" and "soft" varieties—modern investigators prefer to speak of probable trends (perhaps highly probable, but still only probable) rather than inevitable results.

Finally, there is "history," the most problematic term of the three. To begin with, it is not at all self-evident how to characterize any

minism is provided (rather unexpectedly) by Joseph Schumpeter on pp. 129 and 130 of his *Capitalism, Socialism, and Democracy* (second edition; Harper, 1947 [1942]) in a passage worth quoting in full: "Even if mankind were as free to choose as a businessman is free to choose between two competing pieces of machinery, no determined value judgment necessarily follows from the facts and relations between facts that I have tried to convey. . . . Whether favorable or unfavorable, value judgments about capitalist performance are of little interest. For mankind is not free to choose. This is not only because the mass of people are not in a position to compare alternatives rationally and always accept what they are being told. There is a much deeper reason for it. Things economic and social move by their own momentum and the ensuing situations compel individuals and groups to behave in certain ways whatever they may wish to do—not indeed by destroying their freedom of choice but by shaping the choosing mentalities and by narrowing the list of possibilities from which to choose. If this is the quintessence of Marxism then we all of us have got to be Marxists."

9. Raymond Williams, "Base and Superstructure in Marxist Critical Theory," in *Problems in Materialism and Culture: Selected Essays* (Verso, 1980), p. 37.

historical outcome. Consider, for example, the many late-nineteenth-century predictions that the advent of electrical power would lead to a decentralized electronic utopia.[10] Has history proved these predictions (technological cause, historical result) right, or wrong? Wrong: Consider urban skyscrapers, traffic gridlock, and centralized data banks. Right: Consider suburbs, telecommuting, and the decentralization of consumption through domestic appliances. You can also say that technology produces historical contradictions; this may be a more accurate response, but it is evasive if "contradictions" is simply taken to mean a mixed outcome.

In defining "history" in relation to technological determinism, furthermore, we are in danger of going around in logical circles. We tend to do implicitly what Robert Heilbroner does explicitly at the outset of his classic article "Do Machines Make History?" and also in "Technological Determinism Revisited." As Heilbroner stresses, the modern age has defined history in terms of socioeconomic factors rather than in terms of, say, political or diplomatic or religious events. This way of defining history is itself a result of priorities that are technology-based, if not technology-determined. Technology decisively entered the study of history in the Enlightenment and the early nineteenth century. In response to the great technological event of that epoch—the overwhelming and unprecedented increase in productivity—the concept of technological progress was gradually extrapolated to history as a whole, and history became redefined as the record of socioeconomic progress. In other words: For those of us living in the modern age, history is almost by definition a technology-driven process. As Heilbroner so powerfully reminds us, when we talk about how machines make history we must always bear in mind the socioeconomic meaning we attach to the crucial last word.[11]

10. These predictions are ably summarized by James W. Carey and John J. Quirk in "The Mythos of the Electronic Revolution," *American Scholar* 39 (spring 1970): 219–240 and 39 (summer 1970): 395–424.

11. See also Leo Marx's discussion of the role of technology in redefining history, and of the problem of going around in hermeneutic circles, in "The Idea of 'Technology' and the Tenor of Postmodern Pessimism," a paper delivered at the International Workshop on Technological Pessimism, Modern Societies and Their Environments, Tel Aviv and Jerusalem, 1992. (See also Marx's essay in the present volume.)

Even this brief analysis of the words "technology," "determines," and "history" suggests both the benefits and the limitations of semantic analysis. Plainly, we need to unpack these words in order to use critically language that is all too often used uncritically. On the other hand, we may also end up juggling categories, going in linguistic circles, and tossing around abstractions that eventually float free from connections to the social world of material interests and intentions. It would be more in the spirit of Karl Marx (which must hover over any discussion of this subject) to ask why any individual or group would be motivated to assert technological determinism—hard, soft, or otherwise. To paraphrase Heilbroner: We need to think of historical determinism itself in historical and even in political terms.

Interests and motivations are often better revealed by what people deny than by what they assert. To affirm that technology drives history is to deny that God does. Marx in particular was denying that history is directed by God, or by some similar Hegelian or idealistic power. But technological determinists do more than slay God the Father; they also slay Mother Nature, or at least declare her death. Marx was by no means alone in declaring that nature had ceased to be an independent force in history, that its role had been displaced by the "second nature" of human-made artifice. A whole epoch of bourgeois thought is involved here—an epoch in which "technocratic consciousness" (Jürgen Habermas's term) affirms that history is made by humanity as nature is made by God. History is declared to be part of what Sir Francis Bacon called "the human empire." Technological determinism declares humanity's liberation from spiritual and natural necessity alike.

To declare independence is to make a revolutionary political statement—and technological determinism has repeatedly been appealed to as a revolutionary force. Once again, Marx is the obvious example, though one that must be handled with caution. If Marx was a technological determinist at all, he was an exceedingly subtle and ambivalent one.[12] Still, Marx did announce that his mission was to change the world. Paradoxically enough, he proposed that what appears to be determinism actually liberates the proletariat to assume its historical destiny. If capitalist politics or ideology had any significant role in shaping history, then maybe, through some manipulations or

12. See especially the articles by MacKenzie and Bimber cited in note 1.

concessions or new ideas, capitalism could save itself. But since its fatal contradictions arise from its mode of production (and however one wants to interpret that phrase, it is clearly grounded in technology), collapse cannot be averted. Revolution is inevitable precisely because technology is largely out of human control.

The connection between historical theories of technological determinism and the politics of revolution predates Marx, however. The connection was first established by Enlightenment *philosophes*—particularly Anne Robert Jacques Turgot (1727–1781), whose 1750 *Discours sur les progrès successifs de l'esprit humain* is commonly cited as marking the opening of the Enlightenment, and Marie Jean Antoine Nicolas Caritat, marquis de Condorcet (b. 1743), whose suicide in 1793 is commonly cited as marking its end. Turgot's *Discours* and Condorcet's *Esquisse d'un tableau historique des progrès de l'esprit humain* (composed while Condorcet was hiding from the Jacobins) are generally recognized as enormously influential statements of the Enlightenment theory of inevitable historical progress. What is not generally recognized is that they base this theory on a "hard" technological determinism.

In the *Discours,* Turgot outlines the essence of the argument: that historical progress in time is determined by the creation of systems of transportation and communication across space. For Turgot, history is by definition global; it is the story of "the human race," which to the "eye of a philosopher" appears as "one vast whole." Historical progress records the gradual enlightenment of the global human mind. More specifically, Turgot narrates how groups of human beings have become more enlightened in direct proportion to the frequency and intensity of their contacts with other groups. Through most of history, lasting progress was doomed because these contacts were so few and feeble. Genius might appear in one locality—in ancient Athens, to cite a notable example—but would inevitably disappear in the ocean of time. Humankind could innovate, but could never accumulate its innovations.

The inconclusive ebb and flow of history would have continued indefinitely had it not been for the crucial inventions that finally overcame historical entropy and pushed the human race onto the track of cumulative progress. The first and greatest invention was language. In Turgot's words, language "made of all the individual stores of knowledge a common treasure-house, which one generation transmits to another, an inheritance which is always being enlarged by the discoveries of each age." Next, the discovery of writing meant that genius, until then at the mercy of local oblivion, could reach a global audience and therefore become immortal. The climactic invention was the printing press, which rescued the treasures of antiquity from the dust, brought "light to talents which were being wasted in ignorance," and disseminated the discoveries of modern science throughout the globe. Progress in time depends upon diffusion in space, which in turn depends upon crucial inventions.[13]

For Turgot, intellectual, technical, and political revolution are virtually synonymous. The triumph of the bourgeoisie is identified with the triumph of a scientific world view based on universally valid truths. The aristocracy and the clergy, on the other hand, are allied with the regressive forces of error based in localities. To use a Leninist anachronism, Turgot argues that the scientific revolution will never be safe in one country alone; to be secure, the revolution must be global. The construction of technological systems of communication and transportation to disseminate scientific learning is a political act, for these systems are the weapons that will make the triumph of the bourgeoisie inevitable. Technology is revolution.

These political implications, only implied in Turgot's *Discours,* are spelled out much more clearly in Condorcet's *Esquisse.* For Turgot the main danger to progress had been inertia, routine, and passivity. Condorcet defines a far more active and therefore dangerous opponent: the oppressive, mystifying class of priests. From the very first stages of human development, Condorcet asserts, priests had actively opposed human progress, using "crude cunning" to play upon the "credulity" of the dupes who believed them.[14] In Condorcet's vast historical conspiracy theory, intellectual error had been deliberately used by the powerful to maintain their political dominance. Therefore, truth alone could not conquer error; truth allied with technology, however, could. The technological conquest of space had to be achieved before the place-based power of the clergy and their aristocratic allies could be overcome. The ten chapters of Condorcet's *Esquisse* record the major steps in this conquest.

Like Turgot, Condorcet regarded the invention of a written alphabet as the first crucial step. Once spoken language could be reproduced as enduring and transportable signs, the "progress of the human race [was ensured] forever." Still, there were many obstacles

13. Turgot, *Discours,* pp. 41, 57.

14. Condorcet, *Esquisse,* pp. 17–18.

along the way. Like Turgot, Condorcet used the fate of Greek learning as a cautionary tale; that learning was lost for so long, he says, mainly because the Greeks lacked better means of communication. And Condorcet too praised above all the momentous effects of the invention of printing, which had finally and definitively freed the human mind from spatial limitations: "The public opinion that was formed in this way was powerful by virtue of its size, and effective because the forces that created it operated with equal strength on all men at the same time, no matter what distances separated them."[15]

The appeal to technology as a revolutionary force is therefore not particular to Marxism. It is part of a comprehensive view of inevitable historical progress that emerged in the Enlightenment and still endures, though greatly weakened. Technological determinism is an integral part of that theory of progress, according to which technologies of communication and transportation will conquer not just the clergy and the aristocracy but history itself. The long centuries of ebb and flow, of rise and decline, of truth emerging and sinking back into the trackless ocean of time—those centuries are now over. Technical innovation is the decisive factor that has moved history onto an entirely new pathway of unending progress.

We began by analyzing the terminology of technological determinism; then we moved on to consider the intentions and interests of those who have served as powerful advocates of that idea; now we have to consider the intentions and interests of those who acted upon the idea. In the end, we should pay the most attention to those who build determinative technologies—those least responsive to human will, or (in the helpful language of the historian Thomas P. Hughes) those with the most technological momentum. Technological systems can be designed to be highly responsive to human control—or not to be. Human powerlessness can be part of the design. Fate can be engineered. In that case, we should look at those who choose to invest in large, complex technologies, and consider that they may do so quite deliberately in order to *create* technological determinism.

As Hughes reminds us, engineers build values into their designs, and one value that has motivated some prominent engineers is precisely the desire to construct a comprehensive technological system.

15. Ibid., pp. 7, 100.

(Hughes cites Thomas Edison as a prime example.[16]) But there are less aesthetic, more plainly political reasons for constructing comprehensive systems. After all, in most cases engineers themselves do not control the resources needed to create technologies of great scale and complexity. We need to consider not only the inventors and designers but also the sponsors of such technologies.[17] More precise, we need to pay special attention to the ideological, economic, and political motivations of those who control investment in technological systems that, once in place, significantly reduce the range of subsequent human choices.

When revolutionaries are out of power, their appeal to technological determinism is primarily rhetorical, taking the form of the claim that historical inevitability is on their side. In power, however, they are in a position to act upon that claim. They might well be tempted to speed up the pace of historical evolution by sponsoring the construction of technologies they see as determinative.

Such an oscillation between rhetoric and action is evident in the turbulent careers of Turgot and Condorcet. When Turgot delivered his celebrated *Discours,* he was a 23-year-old law student. In subsequent years, however, Turgot attained positions of political power, first as an *intendant* in Limoges and then as minister of finance under Louis XVI. In these offices he commanded resources that allowed him to embark upon ambitious programs of road and canal building. His tenure as Louis's minister of finance was brief, however, precisely because the aristocracy understood so well the revolutionary implications of his policies. After his fall, Turgot returned to advocating, rather than building, technological systems. (In particular, he toyed with inventions that would make possible the cheap reproduction of texts.) Condorcet's career shows a similar pattern. When his mentor Turgot was in power, Condorcet helped design hydraulic experiments to improve canal construction. After Turgot's fall, Condorcet became deeply involved with projects to establish a universal language

16. See, for example, Thomas P. Hughes, "Technological Momentum in History: Hydrogenation in Germany," *Past and Present* 44 (1969): 106–132; "The Electrification of America: The Systems Builders," *Technology and Culture* 20 (January 1979): 124–161.

17. See Ron Westrum's discussion of technological sponsors in *Technologies and Society: The Shaping of People and Things* (Wadsworth, 1991), described by Eric Schatzberg in a review in *Technology and Culture* (33 (April 1992): 392).

for the moral and physical sciences based on the calculus of probabilities.

The urge to ensure historical inevitability by building its technological base is especially evident in the case of Marxism. Marx wrote about technological determinism; Lenin built it in the form of vast projects of electrification, industrialization, and agricultural collectivization. As the Leninist example shows, the motivation for constructing such systems is not only to speed up the forward march of history, but also to prevent any backward slide by making counterrevolution impossible. In order to break the power of the *kulaks,* Lenin created a system of collective farms based on large agricultural machines that could not be used on smaller plots by independent or semi-independent small proprietors. In the same spirit, Turgot and Condorcet had tried to diminish the power of large aristocratic landholders by creating an internal system of canals and roads so small proprietors could sell their grain on the open market.

But revolutionaries who attain political power are by no means the only authorities who have reason to construct determinative technologies. Any elite may use such technologies to achieve political ends. In "Do Machines Make History?" Heilbroner suggested that the concept of technological determinism works best in this historical epoch, when "forces of technical change have been unleashed, but . . . the agencies for the control or guidance of technology are still rudimentary." I am just adding a reminder that some classes profit mightily from the absence of social controls. Ultimately, not machines but people create technological determinism.

The most obvious example of a technology designed to be beyond social control is the nuclear weaponry of the Cold War era. In the words of the essayist Wendell Berry: "We may choose nuclear weaponry as a form of defense, but that is the last of our 'free choices' with regard to nuclear weaponry. By that choice we largely abandon ourselves to terms and results dictated by the nature of nuclear weapons."[18] What Berry fails to emphasize is the political rationale for the seeming irrationality of turning over the power of decision to a technological system. In order for the threat of massive retaliation to be credible, the element of human decision *had* to be removed,

18. Wendell Berry, "Property, Patriotism, and National Defense," in *The Contemporary Essay,* second edition, ed. D. Hall (St. Martin's, 1989), p. 56.

precisely because rational human beings might choose negotiation over war.

The same principle can be used with regard to conventional weapons. In the late summer and the autumn of 1989, President George Bush made a series of decisions that had the net effect of putting half a million American troops in an offensive posture against an opponent who had been issued an unacceptable ultimatum. Those circumstances were constructed to ensure that in January 1990 the United States Senate would choose to declare war with Iraq. During the debate, one senator asked "Are we supposed to go to war simply because one man—the President—makes a series of unilateral decisions that put us in a box—a box that makes that war, to a greater degree, inevitable?"[19] The onset of Operation Desert Storm four days later provided a clear answer to his question.

In a 1963 essay, the cultural critic Lewis Mumford proposed that throughout history there have coexisted two ideal types of technics: democratic and authoritarian. "What I would call democratic technics," he wrote, "is the small scale method of production, resting mainly on human skill and animal energy but always, even when employing machines, remaining under the active direction of the craftsman or the farmer." Whereas democratic technics dates back to the origins of humanity, Mumford argued, authoritarian technics is relatively recent, beginning around the fourth millennium B.C. It was based on a compelling myth of an absolute ruler "whose word was law" and under whom "cosmic powers came down to earth" in this "entirely new kind of theological-technological mass organization." Authoritarian technics depended upon ruthless physical coercion in which human beings were organized into work armies and military armies for purposes of mass construction and mass destruction. Mumford thus interpreted the nuclear policies of the Cold War as a modern example of "human compulsions that . . . date back to a period before even the wheel had been invented"—most notably the compulsion to create systems of total control over physical nature and ultimately over human beings. Authoritarian technics is designed to be determinative, to place power in the technological system itself. In modern times the center of authority lies not in a visible, all-powerful king, but "in the system itself, invisible but omnipresent."

19. Senator John Kerry (Democrat, Massachusetts), *Congressional Record* 137: 7 (January 11, 1991).

"The ultimate end of this technics," wrote Mumford, "is to displace life, or rather, to transfer the attributes of life to the machine and the mechanical collective. . . ."[20]

Mumford's theory is essentially Manichean; the good, "man-centered, relatively weak, but resourceful and durable" technics of democracy is locked in eternal struggle with the "system-centered, immensely powerful, but inherently unstable" technics of authority.[21] Like any grand theory, this one is wide open to criticism. One could argue that small-scale technologies can restrict human freedom, in their own way, as much as large-scale ones. On the other hand, many contemporary engineers would argue that technological systems are not inherently authoritarian but can be designed to be flexible and democratic in their postmodern "robustness." Most arguable is Mumford's conclusion that technological determinism is essentially an illusion, and that we can regain control of our technics simply by casting off the "myth of the machine." While the myths that justify the construction of authoritarian systems may be regarded as empty illusions, the systems themselves are quite real. If the constructions are of sufficient size and scope, they may indeed do what their creators intend—deprive human beings of the liberty to survive outside those systems.

With all that said, however, Mumford's dualistic theory, crude as it is, challenges us to consider the political agendas behind technological development, and in particular to consider the irrational motives of those who build enormous technological systems—in Mumford's strong words, "their paranoid suspicions and animosities and their paranoid claims to unconditional obedience and absolute power."[22] Capitalist rhetoric, in contrast, has always emphasized economic agendas and rational motives. This is the language that Heilbroner uses in "Technological Determinism Revisited" when he links technological development to a stable and systematic acquisitive drive to maximize possibilities for gain. Economics is based on the essentially rational "principle of 'maximizing,'" which is the "mechanism" that translates "a huge variety of stimuli . . . into a few well-defined behavioral vectors." The drive to maximize fortune is the "'rule' of

20. Lewis Mumford, "Authoritarian and Democratic Technics," *Technology and Culture* 5 (January 1964): 2–6.

21. Ibid., p. 2.

22. Ibid., p. 4.

behavior" in market societies that permits us to understand the rules of technological development in a social order that is based on rational, if limited, exchange values.[23]

This analysis begs some important questions—most notably, who decides (individuals? other groups? which groups?) and what criteria they use in deciding. (Defining fortune and the maximization of possibilities may not be so self-evident as it seems.) More generally, however, one must question the validity of the modernist language of physics, rules, stable drives, reason, and predictable behavior. This is familiar capitalist language, of course, but what does it really mean? To what extent does it correspond to human reality, and to what extent is it a word game, a justification of other motives? Mumford raises the possibility that the language of rationality might serve as a cover for political purposes of control and domination—motives that may or may not be congruent with individual goals of acquiring fortune. He also raises the possibility that capitalist elites might not behave in a manner so radically different from elites of the pre-capitalist past.

At the least, we need to consider Mumford's analysis along with the familiar account of economic rationality to understand the motives behind the construction of powerfully determinative technologies. For example, we cannot account for modern military technologies by relying only on a rational model of the acquisition of fortune. Nor does this model go far to illuminate the industrial history of the former Soviet Union and Eastern Europe between 1917 and 1991. If there was ever a modern example of technological development driven by a myth-based, irrational drive to dominate human beings and nonhuman nature alike, this is it. In one of the strangest intellectual twists of our time, it has been Vaclev Havel—playwright, former president of Czechoslovakia, and ardent Heideggerean—who has provided one of the most radical critiques of Soviet-style industrialization. By doing so, he as much as anyone has inherited Lewis Mumford's mantle as the prophet of postmodern technology.

Havel's analysis of the motives behind the construction of coercive technological systems, as well as his advice on how to liberate ourselves from them, echoes Mumford's critique to a remarkable extent. According to Havel, the ecologically disastrous, economically ineffi-

23. Heilbroner, "Technological Determinism Revisited" (in this volume).

cient industry of the Soviet bloc was based on an irrational drive to construct all-encompassing systems of control: "Communism . . . was an attempt . . . to organize all of life according to a single model, and to subject it to central planning and control regardless of whether or not that was what life wanted." Havel asserts that the dominant reflex of modern civilization, whether capitalist or socialist, has been to address all social problems by amassing more scientific knowledge and technological power in order to construct better systems of control. Like Mumford, Havel argues that this response is ultimately futile; instead, "man's attitude to the world must be radically changed." Human beings must come to value democracy over authority, multiplicity over centralization, personal life rather than impersonal systems. "Things must once more be given a chance to present themselves as they are, to be perceived in their individuality. We must see the pluralism of the world. . . ."[24]

There is one other crucial dimension of technical determinism that is omitted from the capitalist account but that is highlighted by Mumford's explanation: gender. Two years after publishing his essay on democratic and authoritarian technics, Mumford returned to his dualistic thesis in another *Technology and Culture* article, "Technics and the Nature of Man." Although the language of the essay, beginning with the title, may be gender-biased by contemporary standards, its content is quite the opposite. To be sure, Mumford does not address issues of gender directly. Instead, his main argument is that culture, not technology, should be considered the prime determinant of human history. In this essay Mumford uses the terms "biotechnics" and "true polytechnics" to refer to technology that is "life-centered, not work-centered or power-centered" because it is directed by cultural values. He then goes on to stress the importance of women's contributions to biotechnics, especially in agriculture and in the vital horticultural discoveries that must have preceded agriculture by many centuries. Mumford also emphasizes the technical significance of the containers primarily associated with women's activities: hearths, pits, houses, pots, sacks, clothes, baskets, and the like. In contrast, the authoritarian "monotechnics" of the Megamachine has been created almost exclusively by and for males—priests, the armed

24. Vaclev Havel, "The End of the Modern Era," *New York Times,* March 1, 1992.

nobility, bureaucrats, and ultimately the king who takes upon himself the cosmic power of the sun-god, "who characteristically created the world out of his own semen without female cooperation." According to Mumford, then, women have been highly significant in shaping democratic biotechnics, but have been routinely excluded from the creation and operation of authoritarian monotechnics.[25]

Mumford thus alerts us to the dissonance between technological determinism and a feminist understanding of history. To put it crudely: Doesn't technological determinism, whether in theory or in practice, represent a predominantly masculine view of human experience? This is true whether we adopt Mumford's political-mythological explanation for the origins of technological determinism or Heilbroner's rational-economic explanation. Mumford proposed that male elites, in their prideful effort to deny the feminine role in bearing and sustaining life, have deliberately created life-denying technological systems. For his part, Heilbroner explains the origins of technological determinism in drives to maximize possibilities and to acquire fortune, which are fundamental rules of human behavior in market economies. But just what do those "rules" mean for women, who through most of history have been restricted by law and custom from acquiring capital and engaging in market relationships? Is this not a "rule" of masculine rather than of human behavior?

Whether the perspective is political or economic, the fascination with technological determinism surely reflects a bias in favor of producers rather than users—the same productivist bias that has promoted general inattention to the role of women in technological history. As we have seen, technological determinism is habitually discussed from the viewpoint of the engineer, the sponsor, the king, or the bureaucrat. Few women have had the opportunity to participate in the creation or the operation of authoritarian technologies (or, to use more neutral language, technologies with considerable momentum). If the issue of technological determinism were examined from the user's point of view, it might look quite different. On the one hand, systems that are planned to be coercive may be considerably less so in reality. Michel de Certeau has reminded us how users can poach, wander, subvert, and otherwise tinker even with technologies intended to be authoritative. On the other hand, tech-

25. Lewis Mumford, "Technics and the Nature of Man," *Technology and Culture* 7 (July 1966): 310–311.

nologies intended to be "liberating" can have unintended or even opposite effects when considered from the user's viewpoint. As Ruth Schwartz Cowan has demonstrated in *More Work for Mother,* the development of "labor-saving devices" had the net effect of increasing the burden of housework for most women.

The point here is not that women have a different mindset about technology, not that they are more democratic or less acquisitive than men, but that they have a different perspective because they have had a significantly different social experience. The far more controversial question is whether women might, on the basis of their different biological experience, have a different perspective on technological determinism. The assertion that technology determines history, it should be recalled, amounts to a denial that Mother Nature does. Would most women be so ready to proclaim the liberation of human history from natural necessity? Might not women, because they experience biological imperatives in a direct and forceful way, be more alert to nature's continuing limitations on human aspirations?

"Nearly all ecofeminists," writes Janet Biehl, ". . . draw on the idea that women have long been 'associated with nature' in Western culture, by males."[26] What is hotly debated is whether this association has some objective validity or is only a social convention. For example, psycho-biological ecofeminists, such as Andree Collard, have argued that women are uniquely ecological beings; this is because their reproductive systems, which make it possible for them to bring forth and nourish life, link them on a profound level with the rest of the living world. Other ecofeminists, such as Ynestra King, are more likely to argue that the identification of woman and nature is a metaphor, not a biological fact. According to King, this identification is "socially constructed . . . an ideology . . . made up by men as a way to sentimentalize and devalue both."[27] In King's striking phrase, "Women have been culture's sacrifice to nature."[28]

26. Janet Biehl, *Rethinking Ecofeminist Politics* (South End Press, 1991).

27. Ynestra King, quoted in Biehl, *Rethinking* (p. 68).

28. Ynestra King, "Healing the Wounds: Feminism, Ecology, and Nature/Culture Dualism," in *Gender/Body/Knowledge,* ed. A. M. Jaggar and S. Bordo (Rutgers University Press, 1989), p. 129. For a fine overview of the woman-nature metaphor and related issues, see Biehl, *Rethinking Ecofeminist Politics,* the introduction and the first chapter of which are excerpted in *Z Magazine* (4, no. 6 (1991): 66–71).

When so much feminist energy has been spent in opposing natural determinism, in protesting that biology is not destiny, I do not want to drag women back to the sacrificial altar of nature. It is hard to see any benefit in overturning a patriarchal technological determinism for a female-oriented but equally crude natural determinism. On the other hand, a position of absolute social construction is, King says, "disembodied." "The logical conclusion," she continues, "is a rationalized, denatured, totally deconstructed person." As King further argues, ecofeminism faces the challenge of overcoming the crude opposition of natural vs. social determinism in order to shape a "new, dialectical way of thinking about our relationship to nature. . . . It is for ecofeminism to interpret the historical significance of the fact that women have been positioned at the biological dividing line where the organic emerges into the social."[29]

Debates among ecofeminists should not be allowed to obscure their essential, common contribution: their insistence that the social remains inextricably linked to the organic. From such an ecofeminist perspective, theories of technological determinism involve a masculine bias because they are positioned so far to the social side of that dividing line. Instead, technology never has taken and never will take command of history as a causal factor independent of nonhuman nature. To be sure, the desire to reach this state of independence is persistent and strong. In Mumford's words, twentieth-century man seeks an entirely new stage of technics, "a radically different condition, in which he will not only have conquered nature but detached himself completely from the organic habitat."[30] Mumford criticizes this goal as seductive but unattainable. In his view, large authoritarian systems are inherently unstable, whereas modest but durable biotechnics accepts the interdependence of human and nonhuman nature.

Concern with this essential interdependence, with the point "where the organic emerges into the social," is not limited to recent ecofeminism. It has also been the concern of a subordinate but significant strand in the Marxist tradition. I still remember sitting in a freshman class in medieval history at Wellesley College, listening raptly as the professor expounded the theory of hydraulic societies of Karl Wittfogel—an erstwhile Marxist who explained the authoritarian character of Oriental societies by their need for complicated water-

29. King, "Healing the Wounds," p. 131.

30. Mumford, "Technics and the Nature of Man," p. 303.

management systems. What made the theory so appealing to a freshman is also its weakness: its simplistic and universalizing determinism. However, it is an environmental determinism rather than a strictly technological one. For Wittfogel, human agricultural practices and conditions of rainfall that are beyond human control combine to determine a social outcome.[31]

In recent years, some Marxist scholars have been picking up where Wittfogel left off. For the most part, they are geographers who seek to reconcile Marxism with their keen awareness of how the nonhuman environment limits and pressures even the most ambitious human-built technologies. "Historical materialism is finally beginning to take its geography seriously," says one of them, David Harvey, who refers to his own approach as "historical-geographical materialism."[32]

In a world of rapidly depleting aquifers, ozone holes, and global warming, any theory of technological determinism that assumes humanity's triumph over natural necessity is plainly unrealistic and simplistic. On the other hand, assertions of humanity's triumph over nature are not entirely mistaken either. Human imperialism is a fact. Our population and our technologies have reached such a scale that they have intertwined with natural systems. Nature may still be a force, but it is no longer an independent one. We human beings have not escaped from nature, but neither has it escaped from us. Is global warming a technical event or a natural one? We can no longer discern the difference. How about *El Niño*? Ozone depletion? Now that we can kill entire lakes and forests with fumes and fires, alter the food chain with oil spills, and scatter radioactivity across Europe, we dwell in an environment where natural and technological processes have merged. This new hybrid environment may not determine history, but it will profoundly and decisively affect the human fate.

31. See the discussion of Wittfogel in Daniel Worster's *Rivers of Empire* (Pantheon, 1985), pp. 22–30.

32. Harvey, *Condition of Postmodernity*, p. 355.

The Idea of "Technology" and Postmodern Pessimism

Leo Marx

Leo Marx holds that the boundless optimism that bolstered the hopes of Americans until the Second World War has dissipated into "widespread social pessimism." The reasons for this change in attitude, according to Marx, are complex. They are to be found in specific technological disasters (Chernobyl and Three Mile Island), in national traumas (the Vietnam War), and more generally in a loss of faith in technology as "the driving force of progress." Marx places this change of expectations in historical context by examining the role of the mechanical arts in the progressive world view and showing how "both the character and the representation of 'technology' changed in the nineteenth century" from discrete, easily identifiable artifacts (such as steam engines) to abstract, scientific, and seemingly neutral systems of production and control. With its "endless reification" in the late nineteenth and early twentieth centuries, the newly refurbished concept of "technology" became invested with a "host of metaphysical properties and potencies" that invited a belief in it as an autonomous agent of social change. By mystifying technology and attributing to it powers that bordered on idolatry, mid-twentieth-century Americans set themselves up for a fall that prepared "the way for an increasingly pessimistic sense of the technological determination of history." Marx concludes that postmodernist criticism, with its ratification of "the idea of the domination of life by large technological systems," perpetuates the credibility of technological determinism.

The factor in the modern situation that is alien to the ancient regime is the machine technology, with its many and wide ramifications.
—Thorstein Veblen (1904)[1]

"Technological Pessimism" and Contemporary History

"Technological pessimism" may be a novel term, but most of us seem to understand what it means.[2] It surely refers to that sense of disappointment, anxiety, even menace, that the idea of "technology" arouses in many people these days. But there also is something paradoxical about the implication that technology is responsible for today's growing social pessimism. The modern era, after all, has been marked by a series of spectacular scientific and technological breakthroughs; consider the astonishing technical innovations of the last century in medicine, chemistry, aviation, electronics, atomic energy, space exploration, and genetic engineering. Isn't it odd, then, to attribute today's widespread gloom to the presumed means of achieving all those advances: an abstract entity called "technology"?

A predictable rejoinder, of course, is that in recent decades that same entity also has been implicated in a spectacular series of disasters: Hiroshima, the nuclear arms race, the American war in Vietnam, Chernobyl, Bhopal, the Exxon oil spill, acid rain, global warming, ozone depletion. Each of these was closely tied to the use or the misuse, the unforeseen consequences or the malfunctions, of relatively new and powerful science-based technologies. Even if we fully credit the technical achievements of modernity, their seemingly destructive social and ecological consequences (or side effects) have been sufficiently conspicuous to account for much of today's "technological pessimism."

One reason we are ambivalent about the effects of technology in general is that it is difficult to be clear about the consequences of particular kinds of technical innovation. Take, for example, modern

1. Thorstein Veblen, *The Theory of Business Enterprise* (Scribner, 1904; Mentor, 1932), p. 144.

2. I don't recall ever seeing the term in print, but I did contribute a paper, ("American Literary Culture and the Fatalistic View of Technology") to a conference on "Technology and Pessimism," sponsored by the College of Engineering at the University of Michigan, in 1979. See Leo Marx, *The Pilot and the Passenger: Essays on Literature, Technology, and Culture in the United States* (Oxford University Press, 1988).

advances in medicine and social hygiene. This is perhaps the most widely admired realm of science-based technological advances; nonetheless, it is often said today that those alleged advances are as much a curse as a blessing. In privileged societies, to be sure, medical progress has curbed or eliminated many diseases, prolonged life, and lowered the death rate; in large parts of the underdeveloped world, however, those very achievements have set off a frightening and possibly catastrophic growth in the population, with all its grim ramifications. Is it any wonder, in view of the plausibility of that gloomy view, that advances in medicine may issue in pessimism as well as optimism?

On reflection, however, such inconclusive assessments seem crude and ahistorical. They suffer from a presentist fallacy like that which casts doubt on the results of much public opinion polling. It is illusory to suppose that we can isolate for analysis the immediate, direct responses to specific innovations. Invariably people's responses to the new—to changes effected by, say, a specific technical innovation—are mediated by older attitudes. Whatever their apparent spontaneity, such responses usually prove to have been shaped by significant meanings, values, and beliefs that stem from the past. A group's responses to an instance of medical progress cannot be understood apart from the historical context, or apart from the expectations generated by the belief that modern technology is the driving force of progress.

Technological Pessimism and the Progressive World Picture

The current surge of "technological pessimism" in advanced societies is closely bound up with the central place accorded to the mechanic arts in the progressive world picture. That image of reality has dominated Western secular thought for some two and a half centuries. Its nucleus was formed around the late-eighteenth century idea that modern history itself is a record of progress. (In the cultures of modernity, conceptions of history serve a function like that served by myths of origin in traditional cultures: They provide the organizing frame, or binding meta-narrative, for the entire belief system.) Much of the extravagant hope generated by the Enlightenment project derived from a trust in the virtually limitless expansion of new knowledge of—and thus enhanced power over—nature. At bottom this historical optimism rested upon a new confidence in humankind's capacity, as exemplified above all by Newtonian physics and the new mechanized motive power, to discover and put to use the essential order—the basic "laws"—of Nature. The expected result was to be a steady, continuous, cumulative improvement in all conditions of life. What requires emphasis here, however, is that advances of science and the practical arts were singled out as the primary, peculiarly efficacious, agent of progress.

In the discourse of the educated elite of the West between 1750 and 1850, the idea of progress often seems to have been exemplified by advances in scientific knowledge; at more popular levels of culture, however, progress more often was exemplified by innovations in the familiar practical arts. Whereas "science" was identified with a body of certain, mathematically verifiable knowledge—abstract, intangible, and recondite—the mechanic arts were associated with the common-sense practicality of everyday artisanal life as represented by tools, instruments, or machines. Nothing provided more tangible, vivid, compelling icons for representing the forward course of history than recent mechanical improvements like the steam engine.[3]

A recognition of the central part that the practical arts were expected to play in carrying out the progressive agenda is essential for an understanding of today's growing sense of technological determinism—and pessimism. The West's dominant belief system, in fact, turned on the idea of technical innovation as a primary agent of progress. Nothing in that Enlightenment world-picture prepared its adherents for the shocking series of twentieth-century disasters linked with—and often seemingly caused by—the new technologies. Quite the contrary. With the increasingly frequent occurrence of these frightening events since Hiroshima, more and more people in the "advanced" societies have had to consider the possibility that the progressive agenda, with its promise of limitless growth and a continuing improvement in the conditions of life for everyone, has not been and perhaps never will be realized.[4] The sudden dashing of

3. Thus, Diderot's *Encyclopedia,* a work that epitomizes Enlightenment wisdom and optimism, is a virtual handbook of technologies, most of them of modern origin.

4. In the United States, politicians like to call the progressive agenda "the American Dream." It is worth noting that in the 1992 election campaign a stock argument of the Democrats was that the current generation may well be "the first whose children are going to be less well off than themselves."

those long-held hopes surely accounts for much of today's widespread technological pessimism.

All of this may be obvious, but it does not provide an adequate historical explanation. To understand why today's social pessimism is so closely bound up with the idea of technology, it is necessary to recognize how both the character and the representation of "technology" changed in the nineteenth century. Of the two major changes, one was primarily material or artifactual; it had to do with the introduction of mechanical (and, later, chemical and electrical) power and with the consequent development of large-scale, complex, hierarchical, centralized systems such as railroads or electric power systems. The second, related development was ideological; it entailed the atrophy of the Enlightenment idea of progress directed toward a more just, republican society, and its gradual replacement by a politically neutral, technocratic idea of progress whose goal was the continuing improvement of technology. But the improvement of technology also came to be seen as the chief agent of change in an increasingly deterministic view of history.

Understanding these changes is complicated, however, by the fact that the most fitting language for describing them came into being as a result of, and indeed largely in response to, these very changes.[5] The crucial case is that of "technology" itself. To be sure, we intuitively account for the currency of the word in its broad modern sense as an obvious reflex of the increasing proliferation, in the nineteenth century, of new and more powerful machinery. But, again, that truism is not an adequate historical explanation. It reveals nothing about the preconditions—the specific conceptual or expressive needs unsatisfied by the previously existing vocabulary—that called forth this new word. Such an inquiry is not trivial, nor is it "merely" semantic. The genesis of this concept, as embodied in its elusive

5. As Raymond Williams famously discovered, this dilemma invariably affects efforts to interpret the cultural transformation bound up with the onset of urban industrial capitalism. His own study turned on five key words ("culture," "industry," "class," "art," and "democracy") whose modern meanings and whose currency derived from the very historical developments he was interpreting. This is, of course, the historical basis for the "hermeneutical circle," which some regard as vitiating all research in the humanities. See the preface to Williams's book *Culture and Society* (Columbia University Press, 1960).

prehistory, is a distinctive feature of the onset of modernity.[6] Not only will it illuminate the rise of "technological pessimism," but it will help us to see that that phenomenon, far from being a direct reflex of recent events, is an outcome of the very developments that called into being, among other salient features of modernity, the idea of "technology."

The Changing Character of the "Mechanic Arts" and the Invention of "Technology"

When the Enlightenment project was being formulated, after 1750, the idea of "technology" in today's broad sense of the word did not yet exist. For another century, more or less, the artifacts, the knowledge, and the practices later to be embraced by "technology" would continue to be thought of as belonging to a special branch of the arts variously known as the "mechanic" (or "practical," or "industrial," or "useful")—as distinct from the "fine" (or "high," or "creative," or "imaginative")—arts. Such terms, built with various adjectival modifiers of "art," then were the nearest available approximations of today's abstract noun "technology"; they referred to the knowledge and practice of the crafts. By comparison with "technology," "the practical arts" and its variants constituted a more limited and limiting, even diminishing, category. If only because it was explicitly designated as one of several subordinate parts of something else, such a specialized branch of art was, as compared with the tacit uniqueness and unity of "technology," inherently belittling. Ever since antiquity, moreover, the habit of separating the practical and the fine arts had served to ratify a set of overlapping and invidious distinctions: between things and ideas, the physical and the mental, the mundane and the ideal, female and male, making and thinking, the work of enslaved and of free men. This derogatory legacy was in some measure erased, or at least masked, by the more abstract, cerebral, neutral word "technology." The term "mechanic arts" calls to mind men with

6. To be sure, the prehistory of all words, perhaps especially all nouns, is elusive, for the investigator must devise ways of referring to that for which adequate names were conspicuously lacking. We need a comprehensive history of the word "technology"—a project that is, or should be, of primary concern to practitioners of that relatively new, specialized branch of historical studies, the history of technology.

soiled hands tinkering with machines at workbenches, whereas "technology" conjures up images of clean, well-educated technicians gazing at dials, instrument panels, or computer monitors.

These changes in the representation of technical practices were made in response to a marked acceleration in the rate of initiating new mechanical or other devices and new ways of organizing work. During the early phase of industrialization (ca. 1780–1850 in England, ca. 1820–1890 in the United States), the manufacturing realm had been represented in popular discourse by images of the latest mechanical inventions: water mill, cotton gin, power loom, spinning jenney, steam engine, steamboat, locomotive, railroad "train of cars," telegraph, factory. The tangible, manifestly practical character of these artifacts matched the central role as chief agent of progress accorded to instrumental rationality and its equipment. Thus the locomotive (or "iron horse") often was invoked to symbolize the capacity of commonsensical, matter-of-fact, verifiable knowledge to harness the energies of nature. It was routinely depicted as a driving force of history. Or, put differently, these new artifacts represented the innovative means of arriving at a socially and politically defined goal. For ardent exponents of the rational Enlightenment, the chief goal was a more just, more peaceful, and less hierarchical republican society based on the consent of the governed.

As this industrial iconography suggests, the mechanic arts were widely viewed as a primary agent of social change. These icons often were invoked with metonymical import to represent an entire class of similar artifacts, such as mechanical inventions; or the replacement of wood by metal construction; or the displacement of human, animal, or other natural energy sources (water or wind) by engines run by mechanized motive power; or some specific, distinctive feature of the era ("the annihilation of space and time," "The Age of Steam,"); or, most inclusive, that feature's general uniqueness (the "Industrial Revolution"). Thus, when Thomas Carlyle announced at the outset of his seminal 1829 essay "Signs of the Times" that, if asked to name the oncoming age, he would call it "The Age of Machinery," he was not merely referring to actual, physical machines, or even to the fact of their proliferation.[7] He had in mind a radically new kind of ensemble typified by, but by no means restricted to, actual mechanical

7. Thomas Carlyle, *Critical and Miscellaneous Essays* (Bedford, Clark & Co., n.d.), III, pp. 5–30.

artifacts. "Machinery," as invoked by Carlyle (and soon after by many others), had both material and ideal (mental) referents; it simultaneously referred to (1) the "mechanical philosophy," an empirical mentality associated with Descartes and Locke and with the new science, notably Newtonian physics; (2) the new practical, or industrial, arts (especially those using mechanized motive power); (3) the systematic division of labor (the workers as cogs in the productive machinery); and (4) a new kind of impersonal, hierarchical, or bureaucratic organization, all of which could be said to exhibit the power of "mechanism." Carlyle's essay is an early, eloquent testimonial to the existence of a semantic void and to the desire to fill it with a more inclusive, scientistic, and distinctive conception of these new human powers than was signified by the most inclusive term then available, "the mechanic arts."

During the nineteenth century, discrete artifacts or machines were replaced, as typical embodiments of the new power, by what later would come to be called "technological systems."[8] It is evident in retrospect that the steam-powered locomotive, probably the nineteenth century's leading image of progress, did not adequately represent the manifold character or the complexity of the mechanic art of transporting persons and goods by steam-powered engines moving wagons over a far-flung network of iron rails. To represent such complexity, that image of a locomotive was no more adequate than the term "mechanic art." As Alfred Chandler and others have argued, the railroad probably was the first of the large-scale, complex, full-fledged technological systems.[9] In addition to the engines and other

8. For the modern concept, see Jacques Ellul, *The Technological System,* J. Neugroschel (Continuum, 1980); for a recent application of the concept to American history, see Thomas P. Hughes, *American Genesis: A Century of Invention and Technological Enthusiasm* (Viking, 1989). But some earlier social theorists who did not use the same term nonetheless anticipated most features of the concept. Few nineteenth-century thinkers devoted more attention to what we call "technology" than Karl Marx; but though he described industrial machinery as embedded in the social relations and the economic organization of an economy dominated by the flow of capital, he still relied, as late as the first (1867) edition of *Capital,* on "machinery," "factory mechanism," and other relics of the old mechanistic lexicon: "In manufacture the workmen are parts of a living mechanism. In the factory we have a lifeless mechanism independent of the workman, who becomes its mere living appendage." (Robert C. Tucker, ed., *The Marx-Engels Reader* (Norton, 1972), pp. 296–297)

9. Rosalind Williams ("Cultural Origins and Environmental Implications of

material equipment (rolling stock, stations, yards, signaling devices, fuel supplies, the network of tracks), a railroad comprised a corporate organization, a large capital investment, and a great many specially trained managers, engineers, telegraphers, conductors, and mechanics. Because a railroad operated over a large geographical area, 24 hours a day, every day of the year, in all kinds of weather, it became necessary to develop an impersonal, expert cohort of professional managers, and to replace the traditional organization of the family-owned and -operated firm with that of the large-scale, centralized, hierarchical, bureaucratic corporation.

Between 1870 and 1920 such large complex systems became a dominant element in the American economy. Although they resembled the railroad in scale, organization, and complexity, many relied on new nonmechanical forms of power. They included the telegraph and telephone network; the new chemical industry; electric light and power grids; and such linked mass-production-and-use systems as the automobile industry (sometimes called the "American" or "Fordist" system), which involved the ancillary production of rubber tires, steel, and glass and which was further linked with the petroleum, highway-construction, and trucking industries. In the era when electrical and chemical power were being introduced, and when these huge systems were replacing discrete artifacts, simple tools, or devices as the char-

Large Technological Systems," *Science in Context* 6 (August 1993): 75–101) argues for a much earlier origin for these systems. She traces their genesis to eighteenth-century Enlightenment philosophers like Turgot and Condorcet, who were committed to the "ideology of circulation." They identified the Enlightenment with the systemic diffusion of ideas and objects in space: ideas through global systems of communication, and objects by means of transportation (road) grids, for the circulation of people and goods. What is not clear, however, is the extent to which circulatory systems of this kind are to be thought of as specifically modern, specifically technological, innovations. After all, the Romans built similarly complex transportation and communication networks. If the point merely is that eighteenth-century theories about circulatory systems anticipated some features—especially the systemic character—of modern technologies, the argument is persuasive. But the systems described in these theories existed primarily in conceptual form, and since they involved no significant material or artifactual innovations it seems misleading to think of them as innovative "technological systems" such as the railroad was. A system is "technological," in my view, only if it includes a significant material or artifactual component. Michel Foucault, who first called attention to these theories of circulation, may have initiated this idealist mode of interpreting their significance.

acteristic material form of the "mechanic arts," the latter term also was being replaced by a new conception: "technology."[10]

The advent of this typically abstract modern concept coincided with the increasing control of the American economy by the great corporations. In Western capitalist societies, indeed, most technological systems (save for state-operated utility and military systems) were the legal property of—were organized as—independently owned corporations for operation within the rules, and for the purposes, of minority ownership. Thus, most of the new technological systems were operated with a view to maximizing economic growth as measured by corporate market share and profitability. At the same time, each corporation presumably was enhancing the nation's collective wealth and power. Alan Trachtenberg has aptly called this fusion of the nation's technological, economic, and political systems "the incorporation of America."[11] By the late nineteenth century, Thorstein Veblen, an exponent of instrumental rationality, ruefully observed that under the regime of large-scale business enterprise the ostensible values of science-based technology (matter-of-fact rationality, efficiency, productivity, precision, conceptual parsimony) were being sacrificed to those of the minority owners: profitability, the display of conspicuous consumption, leisure-class status, and the building of private fortunes. But the abstract, sociologically and politically neutral (one might say neutered) word "technology," with its tacit claim to

10. In explaining the origin of the modern style of corporate management, Alfred D. Chandler describes it has having been "demanded" by the "technological" character of the railroad system. Mechanical complexity, and the consequent need for immense capital investment, were key factors in calling forth a new kind of organization and management. What requires emphasis here, however, is that the effective agent of change in Chandler's widely accepted analysis—the chief cause of the shift, as representative of the technical, from single artifact to system—is the radically new material, or artifactual, character of the railroad. (See Chandler, *The Visible Hand: The Managerial Revolution in American Business* (Harvard University Press, 1977), p. 87ff. and passim.) Historians differ in their accounts of the genesis of the new systems. Thomas Hughes, in *American Genesis* (Viking, 1989) emphasizes changes in modes of organization and management (p. 184ff.), whereas Chandler (whose example, the railroad, belongs to the earlier mechanical phase) emphasizes a material or artifactual change, especially from mechanical to electrical and chemical process.

11. Alan Trachtenberg, *The Incorporation of America* (Hill and Wang, 1982).

being a distinctive, independent mode of thought and practice like "science," is unmarked by a particular socio-economic regime.

Although the English word "technology" (derived from the Greek *teckhne,* "art" or "craft") had been available since the seventeenth century, during most of the next two centuries it had referred specifically and almost exclusively to technical discourses or treatises.[12] In view of the way historians now routinely project the word back into the relatively remote past, it is surprising to discover how recently today's broad sense of "technology" achieved currency. It was seldom used before 1880. Indeed, the founding of the Massachusetts Institute of Technology in 1861 seems to have been a landmark, a halfway station, in its history; however, the *Oxford English Dictionary* cites R. F. Burton's use of "technology" in 1859 to refer to the "practical arts collectively" as the earliest English instance of the inclusive modern usage. (It is important to recognize the exact nature of this change: instead of being used to refer to a written work, such as a treatise, about the practical arts, "technology" now was used to refer directly to the arts—including the actual practice and practitioners—themselves.)

That this broader, modern sense of "technology" was just emerging at the middle of the nineteenth century is further indicated by the fact that Karl Marx[13] and Arnold Toynbee, who were deeply concerned about the changes effected by the new machine power, did not use the word. At points in his influential lectures on the Industrial Revolution (composed in 1880–81) where "technology" would have been apposite, Toynbee, an economic historian, relied on other terms: "mechanical discoveries," "machinery," "mechanical improvements," "mechanical inventions," "factory system."[14] Yet within 20

12. The OED gives 1615 as the date of the word's first use in English, meaning a discourse or treatise on the arts. See "Technology" in the 1955 *Shorter Oxford English Dictionary.* A Harvard professor, Jacob Bigelow, has been credited with anticipating the modern meaning of the word in his 1829 book *Elements of Technology, Taken Chiefly from a Course of Lectures . . . on the Application of the Sciences to the Useful Arts.* (See Dirk J. Struik, *Yankee Science in the Making* (Little, Brown, 1948), pp. 169–170.) Although the scope of the word's meaning has expanded steadily, my impression is that its current meaning retains its essential etymological links to the practical arts and the material world.

13. On Marx's technological vocabulary, see note 8 above.

14. Arnold Toynbee, *The Industrial Revolution* (Beacon, 1960), esp. pp. 63–

years Veblen would be suggesting that the "machine technology" was the distinguishing feature of modernity.[15] My impression is, however, that "technology" in today's singular, inclusive sense did not gain truly wide currency until after World War I, and perhaps not until the Great Depression.

The advent of "technology" as the accepted name for the realm of the instrumental had many ramifications. Its relative abstractness, as compared with "the mechanic arts," had a kind of refining, idealizing, or purifying effect upon our increasingly elaborate contrivances for manipulating the object world, thereby protecting them from Western culture's ancient fear of contamination by physicality and work. An aura of impartial cerebration and rational detachment replaced the sensory associations that formerly had bound the mechanic arts to everyday life, artisanal skills, tools, work, and the egalitarian ethos of the early republic. In recognizing the mastery of various technologies as a legitimate pursuit of higher learning, the universities ratified that shift from the craft ethos of the mechanic arts to the meritocratic aspirations of the engineering and management professions. The lack of sensuous specificity attached to the noun "technology," its bloodless generality, and its common use in the more generalized singular form make the word conducive to a range of reference far beyond that available to the humdrum particularities of "the mechanic arts" or "the industrial arts." Those concrete categories could not simultaneously represent (as either "technology" or, say, "computer technology" can and does) a particular kind of device, a specialized form of theoretical knowledge or expertise, a distinctive mental style, and a unique set of skills and practices.[16]

66. Toynbee was not timid about using new or unconventional terminology, and indeed these lectures were extremely influential in giving currency to the still-novel concept of an "industrial revolution." As late as the eleventh (1911) edition, the *Encyclopaedia Britannica,* which contained no separate entry on "technology," was still offering "technological" as a possible alternative to the preferred "technical" in the entry for "Technical Education" (volume 26, p. 487).

15. See note 1 above. Among other obvious indications that the concept was then in its early, avant-garde stage of development is the way unconventional writers and artists of the period—Herman Melville, Henry Adams, and Oswald Spengler, or the Italian Futurists and Cubists, or (in the next generation) Hart Crane and Charlie Chaplin—used technological images to characterize the distinctiveness of modernity.

16. It is instructive to notice how few of the commonplace statements made

Perhaps the crucial difference is that the concept of "technology," with its wider scope of reference, is less closely identified with—or defined by—its material or artifactual aspect than was "the mechanic arts." This fact comports with the material reality of the large and complex new technological systems, in which the boundary between the intricately interlinked artifactual and other components—conceptual, institutional, human—is blurred and often invisible. When we refer to such systems, as compared with, say, carpentry, pottery, glassmaking, or machine-tool operating, the artifactual aspect is a relatively small part of what comes before the mind. By virtue of its abstractness and inclusiveness, and its capacity to evoke the inextricable interpenetration of (for example) the powers of the computer with the bureaucratic practices of large modern institutions, "technology" (with no specifying adjective) invites endless reification. The concept refers to no specifiable institution, nor does it evoke any distinct associations of place or of persons belonging to any particular nation, ethnic group, race, class, or gender. A common tendency of contemporary discourse, accordingly, is to invest "technology" with a host of metaphysical properties and potencies, thereby making it seem to be a determinate entity, a disembodied autonomous causal agent of social change—of history. Hence the illusion that technology drives history. Of all its attributes, this hospitality to mystification—to technological determinism—may well be the one that has contributed most to postmodern pessimism.

From the Republican to the Technocratic Idea of Progress

As the first complex technological systems were being assembled, and as the new concept of technology was being constructed, a related change was occurring within the ideology of progress. It entailed a subtle redescription of the historical role of the practical arts. Originally, as conceived by such exponents of the radical Enlightenment as Turgot, Condorcet, Paine, Priestley, Franklin, and Jefferson, innovations in science and in the mechanic arts were regarded as necessary yet necessarily insufficient means of achieving general progress.[17] To

nowadays about the import of "technology" actually are applicable to the entire range of existing technologies—medical, military, electronic, domestic, biogenetic, contraceptive, etc.

17. Notice the two distinct uses of "progress" here, each with a markedly

the republican revolutionaries of the Enlightenment (especially the radical *philosophes*), science and the practical arts were instruments of political liberation—tools for arriving at the ideal goal of progress: a more just, more peaceful, and less hierarchical republican society based on the consent of the governed.[18]

The idea of history as a record of progress driven by the application of science-based knowledge was not simply another idea among many. Rather it was a figurative concept lodged at the center of what became, sometime after 1750, the dominant secular world-picture of Western culture. That it was no mere rationale for domination by a privileged bourgeoisie is suggested by the fact that it was as fondly embraced by the hostile critics as by the ardent exponents of industrial capitalism. Marx and Engels, who developed the most systematic, influential, politically sophisticated critique of that regime, were deeply committed to the idea that history is a record of cumulative progress. In their view, the critical factor in human development—the counterpart in human history of Darwinian natural selection in natural history—is the more or less continuous growth of humanity's productive capacity. But of course they added a political stipulation, namely that the proletariat would have to seize state power by revolution if humanity was to realize the universal promise inherent in its growing power over nature. To later followers of Marx and Engels, the most apt name of that power leading to communism, the political goal of progress—of history—is "technology."[19]

But the advent of the concept of technology, and of the organization of complex technological systems, coincided with, and no doubt

different scope of reference: (1) the bounded, internal, verifiable, kind of improvement achievable within a particular practice, such as progress in mathematics, physics, medicine, overland transportation, or textile production; the cumulative effect of such manifold kinds of progress doubtless created the conditions for using the word with a much larger scope of reference: (2) a general improvement in the conditions of life for all of humanity, hence a presumed attribute of the course of events—of history itself.

18. I have summarized this unoriginal interpretation in somewhat greater detail in "Does Improved Technology Mean Progress?" *Technology Review* (January 1987): 33–71.

19. G. M. Cohen, in *Karl Marx's Theory of History* (Clarendon, 1978), makes a strong case for the view that Marx's conception of history was essentially a version of technological determinism.

contributed to, a subtle revision of the ideology of progress. Technology now took on a much grander role in the larger historical scheme—grander, that is, than the role that originally had been assigned to the practical arts. To leaders of the radical Enlightenment like Jefferson and Franklin, the chief value of those arts was in providing the material means of accomplishing what really mattered: the building of a just, republican society. After the successful bourgeois revolutions, however, many citizens, especially the merchants, industrialists, and other relatively privileged people (predominantly white and male, of course), took the new society's ability to reach that political goal for granted. They assumed, not implausibly from their vantages, that the goal already was within relatively easy reach. What now was important, especially from an entrepreneurial viewpoint, was perfecting the means. But the growing scope and integration of the new systems made it increasingly difficult to distinguish between the material (artifactual or technical) and the other organizational (managerial or financial) components of "technology." At this time, accordingly, the simple republican formula for generating progress by directing improved technical means to societal ends was imperceptibly transformed into a quite different technocratic commitment to improving "technology" as the basis and the measure of—as all but constituting—the progress of society. This technocratic idea may be seen as an ultimate, culminating expression of the optimistic, universalist aspirations of Enlightenment rationalism. But it tacitly replaced political aspirations with technical innovation as a primary agent of change, thereby preparing the way for an increasingly pessimistic sense of the technological determination of history.

The cultural modernism of the West in the early twentieth century was permeated by this technocratic spirit. (A distinctive feature of the technocratic mentality is its seemingly boundless, unrestricted, expansive scope—its tendency to break through the presumed boundaries of the instrumental and to dominate any kind of practice.) The technocratic spirit was made manifest in the application of the principles of instrumental rationality, efficiency, order, and control to the behavior of industrial workers. As set forth in the early-twentieth-century theories of Taylorism and Fordism, the standards of efficiency devised for the functioning of parts within machines were applied to the movements of workers in the new large-scale factory system. The technocratic spirit also was carried into the "fine" arts by avant-garde practitioners of various radically innovative styles associated with early modernism. The credo of the Italian Futurists; the vogue of geometric abstractionism exemplified by the work of Mondrian and the exponents of "Machine Art"; the doctrines of the Precisionists and the Constructivists; the celebration of technological functionalism in architecture by Le Corbusier, Mies Van der Rohe, and other exponents of the international style—all these tendencies exemplified the permeation of the culture of modernity by a kind of technocratic utopianism.

Architecture, with its distinctive merging of the aesthetic and the practical, provides a particularly compelling insight into the modern marriage of culture and technology. The International Style featured the use, as building materials, of such unique products of advanced technologies as steel, glass, and reinforced concrete; new technologies also made it possible to construct stripped-down, spare buildings whose functioning depended on still other innovative devices (the elevator, the subway system, air conditioning). This minimalist, functional style of architecture anticipated many features of what probably is the quintessential fantasy of a technocratic paradise: the popular science-fiction vision of life in a spaceship far from Earth, where recycling eliminates all dependence on organic processes and where the self-contained environment is completely under human control.

Postmodern Pessimism

Let us return now to our initial question: how to understand the current surge of technological pessimism. One way to account for the collective despondency, I have suggested, is to chart the advent of the abstract noun that names a quintessentially modern class: "technology." The point is that the idea of a class called "technology," in its ideological inheritance from the practical arts, was suffused from its inception by the extravagant universalist social hopes of the Enlightenment. Those hopes were grounded in what postmodernist skeptics like to call "foundationalism": a faith in the human capacity to gain access to a permanent, timeless foundation for objective, context-free, certain knowledge. The stunning advances of Western science and the practical arts seemed to confirm that epistemological faith, and with it a corresponding belief that henceforth the course of history necessarily would lead to enhanced human well-being.

In their euphoric embrace of that faith, the utopian thinkers of the Enlightenment invented a historical romance called Progress. In it they assigned a heroic role to the mechanic arts. That role, like the romance as a whole, rested on the old foundationalist faith in the capacity of scientific rationalism to yield incontrovertible knowledge. But the part assigned to the mechanic arts in those early years, though heroic, actually was modest compared with what it became once it had been renamed "technology." By the 1920s "technology," no longer confined to its limited role as a mere practical means in the service of political ends, was becoming a flamboyant, overwhelming presence. In many modernist, technocratic interpretations of the romance, "technology" so dominated the action as to put most other players in the wings; in the final act, a happy ending confirmed the vaunted power of technology to realize the dream of Progress. In the aftermath of World War II, however, what had been a dissident minority's disenchantment with this overreaching hero spread to large segments of the population. As the visible effects of technology became more dubious, modernism lost its verve and people found the romance less and less appealing. After the Vietnam era, the ruling theme of Progress came to seem too fantastic, and admirers of the old Enlightenment romance now were drawn to a new kind of post-modern tragicomedy.

Postmodernism is the name given to a sensibility, style, or amorphous viewpoint—a collective mood—made manifest in the early 1970s. As the name suggests, one of its initial motives was a repudiation of the earlier, modernist style in the arts, and a consequent effort to define—and to become—its successor. The successionist aspect of postmodernism was made clear early on by a series of sudden, sharp attacks on modern architecture, probably the most widely recognized of all styles of aesthetic modernism. As early as 1962, in his seminal essay "The Case Against 'Modern Architecture,'" Lewis Mumford, a leading architectural critic and an exponent of early modernism, anticipated many themes of that postmodernist reaction. Most significant here was his analysis of the sources of modernism, an architectural style he traced to a set of preconceptions about the historic role of technology. Chief among them, he wrote, was "the belief in mechanical progress as an end in itself," a belief that rested on the assumption that human improvement would occur "almost automatically" if we would simply devote all our energies to science and technology.[20] As in most of his work, here Mumford's disapproval was not directed at technology in any narrow or intrinsic sense, not at the mere technical or artifactual aspect of modernity, but rather at the larger ideological context; put differently, he was aiming at the imperial domination of architectural practice by the overreaching of the technocratic mentality, whereby the technical means, under the guise of a functionalist style, had become indistinguishable from—and in fact determined—all other aspects of building practice. His target, in sum, was the dominating role of the instrumental in the later, technocratic version of the progressive ideology—a version characteristic of the era of corporate capitalism.

In making this argument, Mumford allied himself with a dissident minority of writers, artists, and intellectuals who had opposed the technocratic idea of progress for a long time. They were adherents, indeed, of a continuously critical, intermittently powerful adversary culture that can be traced back to the "romantic reaction" against the eighteenth-century scientific and industrial revolutions. But the cultural dissidents did not abandon the Enlightenment commitment to the practical arts; what they rejected was the skewed technocratic reinterpretation of that commitment. What they, like Mumford, found most objectionable was the tendency to bypass moral and political goals by treating advances in the technical means as ends in themselves. Nowhere was this criticism made with greater precision, economy, or wit than in Henry David Thoreau's sardonic redescription of the era's boasted modern improvements: "They are but improved means to an unimproved end." So, too, Herman Melville identified a deep psychic root of this warped outlook when he allowed Ahab, the technically competent but morally incapacitated captain of the *Pequod,* a stunning insight into his own pathological behavior: "Now, in his heart, Ahab had some glimpse of this, namely: all my means are sane, my motive and my object mad."[21] As the history of the twentieth century has confirmed, high technical skills may serve to mask, or to displace attention from, the choice of ends. If, as in

20. Donald L. Miller, ed., *The Lewis Mumford Reader* (Pantheon, 1986), p. 75. The essay, which first appeared in *Architectural Record* (131, no. 4 (1962): 155–162), anticipated Robert Venturi's influential 1966 manifesto for postmodernism, *Complexity and Contradiction in Architecture.*

21. See *Walden,* chapter 1; *Moby-Dick,* chapter 41; *The Writings of Thoreau* (Modern Library, 1949), p. 49; *Moby-Dick* (Norton, 1967), p. 161.

Ahab's case, the ends are deformed, amoral, and irrational, such a disjunction of means and ends becomes particularly risky.

This kind of flawed technocratic mentality later became a major target of the radical movement and the counterculture of the 1960s. In retrospect, indeed, that astonishing burst of political outrage looks like a last desperate gasp of Enlightenment idealism. It was an attempt to put technology back into the service of moral and political ends. It is important, if we are to understand the genesis of postmodernism, to recognize that it appeared immediately after the events of May 1968, just as the ardent cultural radicalism of the Vietnam era was collapsing in frustration and disillusionment. Thus, postmodernism embodied, from its birth, a strong current of technological pessimism.[22] It was a pessimism whose distinctive tenor derived from the adversary culture's inability, for all its astonishing success in mobilizing the protest movements of the 1960s, to define and sustain an effective anti-technocratic program of political action.

In conclusion, let me suggest two ways of looking at the technological aspect of postmodern pessimism. For those who continue to adhere to the promise of Enlightenment rationalism, the postmodernist repudiation of that optimistic philosophy is bound to seem pessimistic. Postmodernism not only rejects the romance of Progress; it rejects all meta-narratives that ostensibly embody sweeping interpretations of history. For those who are drawn to the philosophic skepticism of the postmodernists, however, the repudiation of some of the political hopes that ultimately rested on foundationalist metaphysical assumptions need not be taken as wholly pessimistic. Although such a repudiation surely entails a diminished sense of human possibilities, the replacement of the impossibly extravagant

22. American postmodernism is, in my view, most persuasively and attractively represented in the work of Richard Rorty, but he too is more compelling in his skeptical critique of the philosophical mainstream than in his murky anti-realist epistemology. See, esp., *Consequences of Pragmatism* (University of Minnesota Press, 1982), and *Contingency, Irony, and Solidarity* (Cambridge University Press, 1989). For an acute assessment of the political weaknesses inherent in this outlook, see Christopher Norris, *What's Wrong with Postmodernism, Critical Theory and the Ends of Philosophy* (Johns Hopkins University Press, 1990). There have been many efforts to define postmodernism, but perhaps the anti-Enlightenment aspect important here is most clearly set forth by David Harvey (*The Condition of Postmodernity* (Blackwell, 1980)) and Jean-Francois Lyotard (*The Postmodern Condition* (University of Minnesota Press, 1984)).

hopes that had for so long been attached to the idea of "technology" by more plausible, realistic aspirations may, in the long run, be cause for optimism.

But the second way of looking at the role of technology in postmodernist thinking is much less encouraging. What many postmodernist theorists often propose in rejecting the old illusion of historical progress is a redescription of social reality that proves to be even more technocratic than the distorted Enlightenment ideology they reject. Much early postmodernist theorizing took off from a host of speculative notions about the appearance of a wholly unprecedented kind of society, variously called "post-Enlightenment," "post-Marxist," "post-industrial," or "post-historic." A common feature of these theories—and of the umbrella concept, postmodernism—is the decisive role accorded to the new electronic communications technologies. The information or knowledge they are able to generate and to disseminate is said to constitute a distinctively postmodern and increasingly dominant form of capital, a "force of production," and, in effect, a new, dematerialized kind of power. This allegedly is the age of knowledge-based economies.

There are strikingly close affinities between the bold new conceptions of power favored by influential postmodern theorists—I am thinking of Jean-Francois Lyotard and Michel Foucault—and the functioning of large technological systems.[23] Power, as defined by these theories, is dynamic and fluid. Always being moved, exchanged, or transferred, it flows endlessly through society and culture the way blood flows through a circulatory system or information through a communications network. In contrast with the old notion of entrenched power that can be attacked, removed, or replaced, postmodernists envisage forms of power that have no central, single, fixed, discernible, controllable locus. This kind of power is everywhere but nowhere. It typically develops from below, at the lower, local levels, rather than by diffusion from centralized places on high. The best way to understand it, then, is by an ascending analysis that initially focuses on its micro, or capillary, manifestations. The most compelling analogy is with the forthcoming mode of fiber-optic communications, an electronic system that is expected to link all tele-

23. Lyotard, *The Postmodern Condition* and *Le Differends* (University of Minnesota Press, 1984); Michel Foucault, *Power/Knowledge* (Pantheon, 1972).

phonic, television, and computer transmission and reception, and all major databanks, in a single national (and eventually global) network.

This outlook ratifies the idea of the domination of life by large technological systems, by default if not by design. The accompanying mood varies from a sense of pleasurably self-abnegating acquiescence in the inevitable to melancholy resignation or fatalism. In any event, it reflects a further increase in the difficulty, noted earlier, of discerning the boundary between what traditionally had been considered "technology," (the material or artifactual armature, which may be a network of filaments) and the other socio-economic and cultural components of these large complex systems. In many respects postmodernism seems to be a perpetuation of—and an acquiescence in—the continuous aggrandizement of "technology" in its modern, institutionalized, systemic guises. In their hostility to ideologies and collective belief systems, moreover, many postmodernist thinkers relinquish all old-fashioned notions of putting the new systems into the service of a larger political vision of human possibilities.[24] In their view, such visions are inherently dangerous, proto-totalitarian, and to be avoided at all costs. The pessimistic tenor of postmodernism follows from this inevitably diminished sense of human agency. If we entertain the vision of a postmodern society dominated by immense, overlapping, quasi-autonomous technological systems, and if the society must somehow integrate the operation of those systems, becoming in the process a meta-system of systems upon whose continuing ability to function our lives depend, then the idea of postmodern technological pessimism makes sense. It is a fatalistic pessimism, an ambivalent tribute to the determinative power of technology. But again, the "technology" in question is so deeply embedded in other aspects of society that it is all but impossible to separate it from them. Under the circumstances, it might be well to acknowledge how consoling it is to attribute our pessimism to the workings of so elusive an agent of change.

24. This is not true of all postmodern theorists. Rorty reaffirms a traditional liberal perspective, though one whose capacity to provide a theoretical basis for the control of technologically sophisticated multi-national corporations is extremely uncertain.

Rationality versus Contingency in the History of Technology

John M. Staudenmaier

As John Staudenmaier's reference points indicate, this essay originated as a summary comment on the presentations made at the 1989 Dibner Institute/MIT workshop on technological determinism. Since revised and expanded, it points to the history of technology's virtual invisibility not just to the larger public but also to other members of the historical profession. Relatively few general historians are aware of the field, and those who are tend to perceive it as narrowly focused and largely irrelevant to their "mainstream" interests in politics and society. Staudenmaier traces the source of this problem to the uninspired way historians of technology "have handled the question of determinism in the past." Ultimately, he asks practitioners in the field to reconsider their subject matter much as general historians are reconsidering theirs. The issue at hand is the extent to which the older internalist tradition with its whiggish emphasis on priority in invention and its interest in producing master narratives of progress, can be modified and amalgamated with the younger contextual tradition, which, like the so-called new history, is primarily concerned with various political and cultural constituencies in the historical process and with the tensions and conflicts between them. Staudenmaier recognizes the centrality of artifacts to the history of technology and the pressing need to pay attention to them and unpack their complex meanings in light of the new history. While observing that we remain "citizens of the Enlightenment" who "only gingerly explore the domain of the uncertain," he nonetheless is confident that such a synthesis can be made. His is a provocative exploration of the epistemological relationships and psychological attitudes that underlie the shared consciousness of those who currently labor in the field.

The volatile, passionate, sometimes acrimonious debate about historical method published in the June 1989 issue of the *American Historical Review* has much to teach historians of technology. The "old history–new history" battle waged by mainstream historians provides a provocative frame of reference for understanding what is at stake when historians of technology address technological determinism. More immediately interesting, perhaps, is the total absence of historians of technology from the *AHR* debate. None of the *AHR* authors shows any awareness of the history of technology. Technology does not appear, to cite an obvious instance, on Gertrude Himmelfarb's list of "the subjects of the new history," whom she accuses of "'clamoring for' . . . not a place on the periphery of history—that they always had—but at the center, and not intermittently but permanently. What they are all seeking is to be 'mainstreamed' into American history. . . ."[1] Virtual invisibility in the larger historical profession, I will argue, has a great deal to do with how we handled the question of determinism in the past.[2] The history of technology's version of the old-new tension, however, provides an opening to the larger debate that promises to help the subdiscipline and the broader historical profession as well.

Lawrence Levine summarizes much of the *AHR* exchange in the following observation:

> There is one area of historiographical unpredictability, however, with which many historians have not learned to make their peace. This involves not changing interpretations of well-agreed-upon standard events but changing notions of *which events—and which people—should constitute the focus of the historian's study.*[3]

1. Her list: "blacks, women, Chicanos, American Indians, immigrants, families, cities." This and the quote appear in her article "Some Reflections on the New History" (*American Historical Review* 94, no. 3 (1989): 661–670). Even Allan MeGill, with his reference to pumps and dikes and his marvelous Zuider Zee historiographical metaphor, shows no awareness of a subdiscipline dedicated to the study of such things. See his "Recounting the Past: 'Description,' Explanation, and Narrative in Historiography," *American Historical Review* 94, no. 3 (1989): 627–653.

2. The original audience for these observations and my own professional home in the Society for the History of Technology lead me to occasionally address my colleagues as "we."

3. Lawrence W. Levine, "The Unpredictable Past: Reflections on Recent American Historiography," *American Historical Review* 94, no. 3 (1989): 671–679 (my emphasis).

Disagreements over which events and which people are suitable as the focus of historical study abound, as in the confrontation at point blank range between Gertrude Himmelfarb and Joan Wallach Scott about the legitimacy of the new histories or David Harlan's questions (aimed at David Hollinger's contextualism):

> With which texts should intellectual historians concern themselves? Any attempt to privilege a particular set of texts is bound to seem *problematic* these days, given the claims of intertextuality.[4]

"Problematic" indeed! Something in the *AHR* symposium on method stirs the blood.

Still, it is reasonable to suspect that, in the mainstream's received understanding of the history of technology, historians of technology ply their trade more or less undisturbed by Levine's ideologically volatile question of appropriate subject matter. For them the methodological predominance of "the artifact" clarifies and simplifies; artifacts and artifact-events, along with the inventors, entrepreneurs, and engineers who preside over them, constitute our subject matter. Such a perception of the field's internalist innocence would help explain why the work of historians of technology is so often ignored in the historical mainstream.[5]

And rightly so, it may well be argued. Artifacts fascinate because people love to see the work of human minds embodied with the sharp-edged precision of a design that works. Not for nothing does Goethe's devil lure Faust with the promise of the sweet technological project. Rationality, experienced as elegant human creativity, can be powerfully seductive in its own right. Nevertheless, for most historians the clean inner logic of technical pedigrees, tracing minute changes in clock escapements or cast-iron bell designs over time, establishing or at least debating competing claims for inventor rights ("Who was first?"): these thin linear strands, accessible mostly to fellow aficionados, offer dull fare to outsiders. They fail to excite interest precisely because their study seems too neat, too enclosed and abstracted from history's turbulence. The history of technology is invisible, perhaps, because its work continues to be seen as con-

4. David Harlan, "Intellectual History and the Return of Literature," *American Historical Review* 94, no. 3 (1989): 581–609 (my emphasis).

5. For descriptive definitions for "internalism," "externalism," and "contextualism" as typically understood in the field, see below.

stricted within the black box of system designs. For the profession at large, the historians of technology reside right alongside the gears and circuits.

Despite the received impression, however, it is precisely in making sense of the artifact's role in historical events that the history of technology has most to say to the larger historical profession. It is here precisely that the subdiscipline engages in the old history–new history conflict. Indeed, Scott's confrontation with Himmelfarb about the old and new historical approaches can serve as defining the frame of reference for arguments about the artifact in the history of technology.

Himmelfarb criticizes "the new history" for "standing outside *the received opinion*—the opinion of contemporaries as well as traditional historians—and [being] prepared to pronounce it simply false." "The old history," she continues, "stands within the received opinion, trying to understand it as contemporaries did, to find out why they believed what they did, why those beliefs seemed to them 'credible.'"[6] Scott challenges the legitimacy of any reading of the historical record in which some consensus stands as *the* true story and in which the stories of those who fought that consensus are ignored or seriously marginalized. Calling such consensus stories "master narratives," she argues:

> Any master narrative—the single story of the rise of American democracy or Western civilization—is shown to be not only incomplete but impossible of completion in the terms it has been written. For those master narratives have been based on *the forcible exclusion of Others' stories.* They are justifications through teleology of the *outcomes* of political struggles, stories which in their telling legitimize the actions of those who have shaped laws, constitutions, and governments—"*official stories.*"[7]

By telling the story of a consensus, and avoiding the tragedies, nobilities, and follies of conflict, the historian implies that things inevitably turned out as they did because the inherent rationality of events ordained that they would.

But nowhere—and now we get to the heart of the matter—can we find a master narrative so deeply entrenched in popular imagination and popular language as in the mythic idea of progress, particularly

6. Gertrude Himmelfarb, *AHR* Forum: "Some Reflections on the New History," *American Historical Review* 94, no. 3 (1989): 669–670.

7. Joan Wallach Scott, "History in Crisis? The Others' Side of the Story," *American Historical Review* 94, no. 3 (1989): 689–690 (my emphasis).

technological progress. Already in 1974, Reinhard Rürup articulated the problem for historians of technology in language strikingly similar to Scott's 1989 passage cited above:

> All too often the current fruits of technology appeared to be the *necessary outcome* of previous inventions and techniques, and one was inclined in retrospect to assume that from the very beginning the great inventions carried, as it were, the seed of all subsequent related developments, a seed which only awaited germination according to some "natural law" of technology. What this approach lacked above all—and for obvious reasons—was critical detachment from the object of research; in a sense it was "*company history.*"[8]

Scott and Rürup address precisely the same problem. Both criticize historical accounts that settle for what they understand to be an excessively tidy rationality, screening out failed alternatives and rendering mute those constituencies who rejected the position that eventually won out.

Rürup's position is hardly unique. For three decades historians of technology have condemned the "company history" approach repeatedly. The master narrative they find so distasteful (sometimes called "technological determinism") rests on the radically ahistorical premise that there exists a quasi-preternatural entity called "Technology." Since "Technology" is defined as the application of modern (Western) science, and since Science is understood to operate free from all bias in its pursuit of objective truth, technology must not be impeded by Luddite romanticism from continuing the triumphant march by which Western societies conquer nature.

But if there exists an entity one can name "Technology" (as opposed to many distinct "technologies"), and if it advances with inevitable necessity along a predetermined path, then there is no point whatever to serious study of the historical particulars that constitute the context of change for any single technology. Historians of technology find such pure whig history repugnant because it puts them out of business unless they are willing to sign on as public-relations hucksters—image enhancers—for the captains of progress.[9]

8. Reinhard Rürup, "Historians and Modern Technology: Reflections on the Development and Current Problems of the History of Technology," *Technology and Culture* 15, no. 2 (1974): 174–175 (my emphasis).

9. I use the common expression "whig history" for historical interpretation that treats its central subject matter (in my own profession, the technological artifact) in abstraction from its ambience. For the classic discussion of the

Why, then, has such an ahistorical brand of determinism continued to evoke such concern in the field? Why convene a conference at MIT, and publish its collected works, for one more charge up this same historiographical and ideological hill? Let me suggest two related reasons. First, historians of technology find some aspects of their largely internalist origins hard to reconcile with the repugnance they feel for the myth of autonomous progress. Second, progress ideology has an extraordinary and mostly underestimated hold on popular rhetoric in American culture. Let us consider both briefly.

First, the Society for the History of Technology (SHOT) was formally organized in 1958, when "history" and "consensus" were almost synonyms, by a group of people with close ties to the engineering community. They began as dissenters from the History of Science Society, where they had grown accustomed to and rebellious about the whiggish definition of technology as applied science.[10] With a few prominent exceptions, such as Lewis Mumford, Louis Hunter, Lynn White, Jr., their best-recognized intellectual heritage included a half-millennium's hardware catalogues and treatises and a half-century's pious celebrations of engineers, inventors, and their marvelous works. Historians of technology began to organize themselves just as Singer, Holmyard, Hall, and Williams's five-volume *History of Technology* and several equally internalistic multi-volume histories of technology were appearing in print.[11] These histories represented monumental efforts to catalogue and organize many decades' worth

term see Herbert Butterfield's *The Whig Interpretation of History* (G. Bell & Sons, 1951). For a discussion of the technological-determinism debate as found in the first 20 years of *Technology and Culture* see pp. 134–148 of my book *Technology's Storytellers: Reweaving the Human Fabric* (MIT Press, 1985).

10. Philip Scranton has called historians of technology "lapsed (in the theological sense) scientists and engineers" ("None-too-Porous Boundaries: Labor History and the History of Technology," *Technology and Culture* 29, no. 4 (1988): 731). See also Brooke Hindle, "A Retrospective View of Science, Technology, and Material Culture in Early American History," *William and Mary Quarterly* 11, no. 1 (1984): 427, and chapter 1 of *Technology's Storytellers.*

11. Charles E. J. Singer, E. J. Holmyard, A. R. Hall, and Trevor I. Williams, eds., *A History of Technology,* five volumes (Oxford University Press, 1954–1958); Maurice Daumas, ed., *Histoire Generale des Techniques,* three volumes (Presses Universitaires de France, 1962–1968); A. Zvorikine et al., eds., *Geschichte der Technik,* two volumes (Leipzig, 1964; originally published in USSR, 1962). For a more detailed discussion of the origins and ancestry of the Society for the History of Technology, see pp. 1–12 of *Technology's Storytellers.*

of scholarly and antiquarian attention to the evolving designs of the artifacts that had come to be included in the canon of significant technologies. Against the weight of this long- and well-worked heritage of internalist scholarship, SHOT's three decades of history-making looks more like a beginning than like a mature flowering of a non-whiggish alternative reading of technological change.

Second, and very likely more significant, the place of the concept "Technology" in popular consciousness continues to make the ideological and intellectual boundaries of the history of technology exceedingly difficult to identify or establish. Because of the peculiar nature of the West's love affair with progress and its identification of progress with science and technology, to say nothing of the marketing importance of an upbeat image for current technologies, the question of who gets to tell the story of technological change has been and continues to be hotly contested by an ideologically diverse array of individuals and institutions.[12] Consider some of the claimants.

O. B. Hardison, Witold Rybczynski, and a host of others who are certainly more reputable than company-history hacks but who are just as certainly not historians of technology, publish a never-ending stream of books and articles addressing one or another aspect of the "Technology" question.[13] From another perspective, the United States' premier technical museums reflect the pressures that attend technological storytelling. The Smithsonian Institution's National Museum of American History and the Henry Ford Museum–Greenfield Village complex stand alone among American museums for the richness of their collections and the sophistication of their curatorial personnel. Both have recently mounted multi-million-dollar exhibits. The Smithsonian's "Engines of Change" attempts, in its printed support materials, to balance the story of nineteenth-century American industrialization with a mix of positive and negative interpretations.

12. For a more thorough interpretation of progress ideology in the United States see Christopher Lasch's "The Idea of Progress in Our Time" and my "Perils of Progress Talk: Some Historical Considerations" in *Science, Technology and Social Progress,* ed. S. Goldman (Lehigh University Press, 1989).

13. For a few recent examples see O. B. Hardison's *Disappearing Through the Skylight: Culture and Technology in the Twentieth Century* (Viking, 1989) and Witold Rybczynski's *Taming the Tiger: The Struggle to Control Technology* (Viking Press, 1983). James Burke's still-popular television series "Connections" exemplifies an approach that is almost completely innocent of what most historians of technology would call a critical historical perspective.

On the sensual level, however, the aesthetic splendor of precision machinery dominates so completely that most visitors will perceive the story as unalloyed good news. Sadly, the Smithsonian's more recent exhibit "Information Age: People, Information and Technology" suffers from a strain of the progress virus a good bit more virulent than the one afflicting "Engines of Change." The Ford Museum's spectacular exhibit "The Automobile in American Life" achieves a marvelous evocation of America's love affair with cars, but it too has difficulty portraying such recent problems as pollution, gridlock, and highway-maintenance costs.[14] Finally, consider unvarnished corporate hype. The custom of inviting major corporations to install pavilions at World's Fairs, which began with the Century of Progress Exposition (Chicago, 1933), continues today at the extraordinarily popular EPCOT (Experimental Prototypical City Of Tomorrow) Center at Disney World. Here firms such as General Motors, Exxon, and General Electric devote substantial resources toward teaching an acceptable version of the history of their technology. EPCOT attracts 15 million to 20 million patrons every year, each willing to endure long lines and high prices to be indoctrinated in a relentlessly smiling contempt for past technological achievements together and with faith in a sanitized, inexorably beneficial, technological future. I have yet to meet a historian of technology who left EPCOT untroubled by the doctrinaire catechetical extremes of the Disney imagineers. Still, some non-professionals whom I consider wise and mature adults love EPCOT. Their response to a Disney-style version of progress reminds us of something that historians of technology can readily overlook: EPCOT succeeds because it evokes profound cultural feelings.[15]

Historians of technology, therefore, conduct their research, teach their classes, and interact with their colleagues in a social context that

14. See my reviews of "Engines of Change" (*Journal of American History* 77, no. 1 (1990): 217–221) and "The Automobile in American Life" (*Public Historian*) 10, no. 3 (1988): 89–92).

15. For a detailed study of progress ideology, its origins, and its modes of operation in public discourse, see my "Perils of Progress Talk: Some Historical Considerations," in *Science, Technology and Social Progress,* ed. Goodman. For a critique of EPCOT see Michael L. Smith's "Back to the Future: EPCOT, Camelot, and the History of Technology," in *New Perspectives on Technology and American Culture,* ed. B. Sinclair (American Philosophical Society, 1986).

is deeply biased, on both the popular and the academic level, toward the very whig history that vitiates their profession. More disturbing still, the master narrative of technological progress operates as a bias within the profession. The determinism question keeps coming back to life because historians of technology find it as hard to exorcise its Western chauvinism and its seductive promise of sweet rationality from themselves as to face into powerful deterministic prevailing headwinds generated by the culture at large.

Nevertheless, by the end of SHOT's first decade the members had served notice to one another—and to whomever else might have been listening—that the society was slowly moving away from its internalist "hardware" identity toward a loosely defined but unmistakably contextual methodology. A contextual approach to the history of technology treats the values, biases, motives, and world views of design elites as important evidence for interpreting why a given technical design turned out as it did. Thus, contextual historians have increasingly come to see technologies as expressing political and cultural concerns, and not merely serving functional ends.

The most helpful overview of the history of technology is provided by *Technology and Culture,* SHOT's journal of record. If we consider the entire corpus of *T&C*'s articles as a single subdisciplinary text, we find that, with few exceptions, the research patterns established in SHOT's first 10 years have remained remarkably stable ever since.

Almost from the start *T&C* took on the methodological profile it still maintains.[16] Internalist articles—those treating specific hardware details, with little or no reference to the larger nontechnical context—have maintained a modest but still substantial place in the overall mix. Until 1981 they represented 18% of all articles. In the years 1982–1990 their share fell to a still noteworthy 10%. Externalist articles—those interpreting technological ambience with no specific attention to technical design—have maintained a similar position; since 1967 they have fluctuated only slightly between 16% and 18%.

Contextual articles, which attempt to integrate the design of a technology with some aspect(s) of its context, have steadily expanded

16. I discuss the taxonomic definitions used to score *T&C* articles as "internalist," "externalist," or "contextual" on pp. 202–209 of *Technology's Storytellers.* Note that the following discussion summarizes a more complex analysis (including nonhistorical and historiographical articles), so the percentages don't add up to 100.

their dominant position. In the first seven years they represented 41% of all articles. In the next decade and a half that share climbed gradually to nearly 60%. In the 1980s nearly 70% of *T&C*'s articles were contextual.[17] Still, the contextualist commitment runs deeper than this profile suggests. Both the internalist and the externalist approach to technological history reveal an affinity for the contextual majority position. A significant number of contextual articles are the work of internalists who have learned to appreciate (or so it has seemed to this reader) the importance of non-artifactual contextual detail. In such articles one sees the internalist's affection for the artifact held in tension with respect for the contingencies of historical context. The vast majority of the profession recognizes that the work of understanding precisely how artifacts work and how their designs change is never finished. Internalist research constitutes one essential element in the profession's contextual equation, and it is respected as such. The very existence of externalist research, of course, presupposes that contextual insight that technical artifacts reveal their historical meaning only when understood *in context.* Indeed, externalist research depends on the corpus of internalist and contextual studies to create the artifactual core that their research presumes.[18]

17. It would, of course, be simple-minded to assume that all contextual articles represent a single coherent approach to the relationship between technical design and cultural context. "Contextual" articles in *Technology and Culture* range from such relatively modest and mostly internalist forays across the design–context divide as Robert B. Gordon's study of skilled metalworkers in the mid-nineteenth-century federal small-arms industry ("Who Turned the Mechanical Ideal into Mechanical Reality?"; 29, no. 4 (1988): 744–778) to Joel A. Tarr, James McCurley III, Francis C. McMichael, and Terry Yosie's much more thorough integration of the designs of various waste-water sewage methods with shifting social and fiscal priorities in U.S. cities ("Water and Wastes: A Retrospective Assessment of Wastewater Technology in the United States, 1800–1932," 25, no. 2 (1984): 226–263).

18. One example of the internalist approach in its pure form is William Rostoker, Bennet Bronson, and James Dvorak's "The Cast-Iron Bells of China," *T&C* 25, no. 4 (1984): 750–767. On the other hand, George Ovitt's revisionist critique of Lynn White's famous hypothesis ("The Historical Roots of Our Ecological Crisis," in *Dynamo and Virgin Reconsidered,* ed. L. White, Jr. (MIT Press, 1968)) exemplifies a purely externalist study with no direct reference to any technological design. See Ovitt's "The Cultural Context of Western Technology: Early Christian Attitudes Toward Manual Labor," *T&C* 27, no. 3 (1986): 477–500.

Having considered the tensions found in the history of technology, the current status of whig history, and SHOT's antipathy to the ideology of technological determinism and its gradual shift toward contextual methodology, what can we learn from the papers presented and the discussions pursued during this MIT-Dibner workshop? I will make two observations and then hazard an interpretation.

First, the workshop's rhetoric reincarnated almost every detail of SHOT's 30-year-long determinism debate. Let me oversimplify a little by suggesting that the group divided itself into two groups, somewhat along generational lines.[19] On one side we find people comfortable with rationality, or what Thomas Hughes described, not very favorably, as "a tightly coupled systemic universe." On the opposite side we find people who are comfortable with complexity, ambiguity, conflict, and unresolved issues. The first group says that progress is real, Western, and a very good thing. Its members pay attention to quantified measures such as increasing standards of living and longer life expectancies. They speak a lot about the inherent human desire to control that which is not now docile to human purposes.[20] On the other side we find people who are more at home talking about technology in terms of symbol and power. They ask who creates the symbols that constitute the pervasive, multivalent, and ambiguous moral presence of the various dominant technologies resident in Western societies. For them ambiguity is not a bad thing, unresolved questions are healthy, and differences between various actors (and victims) in technological stories constitute the stuff of the historical record.

19. What follows necessarily oversimplifies both the complexity of any single scholar's position and the amount of diversity among those who could be characterized as members of one or the other "group." I choose to risk the oversimplification because I find the tension between rationality and ambiguity to be a very helpful heuristic model for understanding the difficulty of the profession's task of articulating the place of the artifact in history. When I originally presented these remarks in oral form as one of the concluding "observers" at the MIT-Dibner Conference, I was also calling attention to what I perceived to be a very important tonal characteristic of the discussion.

20. James R. Beniger's *The Control Revolution: Technological and Economic Origins of the Information Society* (Harvard University Press, 1986) represents this view in its pure form. Beniger sees the search for systemic control as the central shaping dynamic of Western culture for the past millennium.

In this workshop, the first group reiterated their confidence in the gradual triumph of Western rationality over nature's constraints and the incremental advance of scientific and technological progress over time. They showed concern that this essentially positive story of science and technology might fade from memory if too much attention were paid to technology's admittedly real "dark side." The second group argued for a reading of the disorderly, sometimes technically irrational dimensions of design decisions as politically or culturally motivated, and of the concept of progress in particular as a conceptual tool by which technical elites try to dominate their inferiors.[21] We could call the first group "the rationality people" and the second group "the ambiguity people."

Second, I was struck by how little direct confrontation occurred between the proponents of what appear to be two wildly different views. Despite the obvious parallel between these disagreements over a Western technological master narrative of progress, the participants did not exhibit the acrimony found in the pages of the *AHR*. Whence this politeness?

Let me hazard a hypothetical fantasy. Imagine the rationality people saying to themselves "Ignore the iconoclasts; we own the store anyway." And imagine the ambiguity people saying to themselves "Ignore the true believers; they are irrelevant and on their way out." It may be, however, that the ambiguity people underestimate the power of the ideology of autonomous progress—not only its symbolic power at work in the larger culture, but also its conceptual power as an interpretative hypothesis. If this hunch comes at all close, then we (I include myself among the ambiguity people) have been polite because we feel that we are not required to debate the merits of the idea. We don't have to confront our counterparts here about their claim that undeniable progress has occurred in the West, that the achievements of rational control must be dealt with, that "tightly coupled systems" have come to dominate the planet. We can evade

21. Presenters at the 1989 Dibner Workshop (in order of presentation): Merritt Roe Smith, Thomas P. Hughes, William H. McNeill, Thomas J. Misa, Rosalind H. Williams, Victor Weisskopf, I. Bernard Cohen, James Bartholomew, Nicolaas Bloembergen, Carl Kaysen, Robert L. Heilbroner, Alfred D. Chandler, Jr., Peter Perdue, John K. Smith, Alan Trachtenberg, Leo Marx, Richard Bulliet, Bruce Mazlish, Robert Howard, Philip Khoury, Colleen A. Dunlavy, Philip Scranton, Michael L. Smith, Bruce Bimber, Gerald Holton, Jill K. Conway, John M. Staudenmaier, Kenneth Keniston.

such discussion by saying that their position is primitive because it ignores its own political and symbolic baggage. But if we feel we can walk away from those "true believers," does that not imply that we claim a sanitized, untainted, and ultimately progressivist position for ourselves?

The ideology of progress, we seem to be saying, contaminates advertisements, structures of government and corporate policy, the EPCOT Center, and our colleagues, but not us. It works its evils only in the unwashed non-elites just named. Such a claim, it should be clear, immediately defines us as victims of the precise ideology we criticize. We would be saying that we are dispassionately rational critics of our own society, as if we ourselves did not bear the cultural valences to which we must obviously stand heir—that we do our historical work outside of history.[22] If this were so, then from our side this workshop's polite and dispassionate discourse would be very appropriate; after all, we discuss what is "out there" in the culture, not what is "in here" in our bellies. From this perspective, then, we see the rationality people as not worth debating because we suffer from the illusion that we have managed to transcend and escape our own culture's view of progressive rationality.

On the other side, one might suggest that the people who identify with Western, quantified, rational, scientific-technological progress reveal an elitism more unblushing than the subtler and more evasive elitism of the ambiguity people. They might be saying "Face it: Western science, Western practice, Western economics have in fact swept the opposition from the field, and the only people who really don't get it are those who are illiterate in the new language of the past couple of hundred years—the language of precisely quantified rationality." At least this brand of elitism sits right there on the table. It is

22. Hans-Georg Gadamer's articulation of the principle in *Wahrheit und Methode* (Mohr, 1960) remains a classic: "A person who imagines that he is free of prejudice, basing his knowledge on the objectivity of his procedures and denying that he is himself influenced by historical circumstance, experiences the power of the prejudices that unconsciously dominate him as a *vis a tergo*. A person who does not accept that he is dominated by prejudices will fail to see what is shown by their light. . . . Historical consciousness in seeking to understand tradition must not rely on the critical method with which it approaches its sources, as if this preserved it from mixing in its own judgments and prejudices. It must, in fact, take account of its own historicality." (*Truth and Method*, tr. G. Barden and J. Cumming (Seabury, 1975), p. 324)

blunt, and, for me, infuriating. I have many friends who would find that perception close to obscene because they approach such questions from a non-Western starting point. Still, what has been striking me as I have reflected on our discourse at this workshop has been my own style of elitist snobbery, which I have been calling the ambiguous-technology position. If I am correct in this reading of our discourse, then, we may be more polite with each other than is good for us. We need to risk revealing our own beliefs and the commitments that generate our passion for telling technological stories.

It will not do, however, to observe that we are too nice to one another and let it go at that. From a different perspective, one as important as the hypothesis of contrasting elitisms I have just advanced, our mutual conviviality is good news rather than bad. Let me suggest that we are also polite in our disagreements because, at bottom, we like one another and we share a kindred interest in technological change. The bond of affection that holds us together has two related elements: First, we know that we work in a highly charged and contested environment, and that not all who tell technological stories respect serious historical scholarship. We share a sense of powerlessness in the face of massive, well-funded, symbol-constructing projects, such as EPCOT, which all of us find distasteful and sometimes heinous. Second, we share affection and respect for human artifacts and the processes by which they are designed and operated.

The epistemological relationships between contextual, internalist, and externalist styles of research outlined above underscore a key dimension in the shared consciousness of historians of technology. Despite some old history–new history differences, few of us (if any) would dispute the profession's requirement that historians of technology respect the internal design constraints that tell us how things work. It bothers historians of technology to see a historical landscape littered with technological black boxes whose inner workings (and the historical significance of those inner workings) remain opaque. Whether motivated by unabashed delight in human creativity or by a passion to identify which elite's interests have been served by a given technology, historians of technology share a common commitment. They try to open the black boxes, to demythologize the ideology of autonomous progress that would render such detailed attention futile, and to restore the essential humanity of the design process. So one could explain our convivial tone by remembering that we have all grown to respect artifacts as critically important dimensions of the historical record.

It seems safe to say that historians of technology experience their academic calling much as other historians do. We remain citizens of the Enlightenment, loving the rational clarity of a well-drawn argument and seeing that which resists clarification as something to be overcome. We only gingerly explore the domain of the uncertain. Gertrude Himmelfarb warns against abandoning the traditional account for some advocacy historian's newly imposed master narrative. She and others fear that the result of the new history might turn out to be the collapse of meaningful conversation about history, a pluralism degenerated into warring camps which respect no diplomacy and take no prisoners. For her part, Joan W. Scott calls for the enfranchisement of stories experienced by the "Others" and trusts the future to the uncertain prospects of multiple perspectives. For her, the threat of ideological monopoly outweighs the danger of pluralistic chaos.

As we have seen, historians of technology wrestle with the same tensions. Rationality versus uncertainty, the coherent narrative versus the unfinished tangle of diversity—such questions remain the stuff of the historical profession. These same tensions define the place of the artifact in history. Historians of technology remind us that artifacts merit our attention, despite their sometimes forbidding black-box complexity, not only because their built-in human biases exert influence on society as long as they continue to be used but also because attention to artifacts helps us to recognize and savor our common human condition. Technological artifacts exist as crystallized moments of past human vision, each one a little "master narrative" seeking to enforce its perspective, each one buffeted by the swirl of passion, contention, celebration, grief, and violence that makes up the human condition.[23] As we pay attention to their history, unpacking the black boxes and restoring the essential humanity of the technical design process, these commitments unite us at a deeper level than our diverse readings of rationality and ambiguity.

23. For one exploration of the complex relationships between artifacts and their environment, see Thomas P. Hughes's essay "Technological Momentum" in this volume.

Contributors

Bruce Bimber is Assistant Professor of Political Science at the University of California at Santa Barbara. He has written on decision making in political institutions and on the politics of expertise.

Richard Bulliet is Professor of History and Director of the Middle East Institute at Columbia University. His book *The Camel and the Wheel* received the Society for the History of Technology's Dexter Prize in 1977.

Robert Heilbroner is Norman Thomas Professor Emeritus at the New School for Social Research. His most recent book is *Twenty-First Century Capitalism.*

Thomas P. Hughes is Andrew W. Mellon Professor of History and Sociology of Science at the University of Pennsylvania. His books include *Networks of Power: Electrification of Western Society, 1880–1930* (winner of the 1984 Dexter Prize) and *American Genesis: A Century of Invention and Technological Enthusiasm.*

Leo Marx is Senior Lecturer and Kenan Professor of American Cultural History Emeritus in MIT's Program in Science, Technology, and Society. His work examines the relationship between technology and culture in nineteenth- and twentieth-century America, and his most recent book is *The Pilot and the Passenger: Essays on Literature, Technology, and Society in America.*

Thomas J. Misa is Associate Professor of History at the Illinois Institute of Technology. He is finishing work on a book examining technological and social change in the context of the American steel industry between 1865 and 1925.

Peter C. Perdue is Associate Professor of History and head of the history faculty at MIT. He is the author of *Exhausting the Earth: State and Peasant in Hunan, 1500–1850.*

Philip Scranton is Professor of History at Rutgers University, and director of the Center for the History of Business, Technology, and Society at the Hagley Museum and Library. His publications include *Figured Tapestry: Production, Markets, and Power in Philadelphia Textiles, 1885–1941.*

Merritt Roe Smith is Cutten Professor of History of Technology and director of the Program in Science, Technology, and Society at MIT. His book *Harpers Ferry Armory and the New Technology* received the Pfizer Award from the History of Science Society and the Frederick Jackson Turner Award from the Organization of American Historians.

Michael L. Smith is Professor of History at the University of California at Davis. His work has addressed the cultural dimensions of environmental sciences, nuclear power, space technology, EPCOT Center, the idea of progress in Cold War America, and gender and technology in postwar advertising. His most recent essay, "Making Time: Representations of Technology at the 1964 World's Fair," appeared in *The Power of Culture,* edited by Richard W. Fox and T. J. Jackson Lears.

John M. Staudenmaier, S.J. is Professor of History at the University of Detroit Mercy. The author of *Technology's Storytellers,* he is working on a book tentatively entitled *Henry Ford: Symbol and Symbol Maker.*

Rosalind Williams is the Robert M. Metcalfe Associate Professor of Writing in the Program in Writing and Humanistic Studies at MIT. She is the author of *Dreamworlds: Mass Consumption in Late Nineteenth-Century France* and *Notes on the Underground: An Essay on Technology, Society, and the Imagination.*

Index